COMPUTATIONAL
METHODS OF
LINEAR ALGEBRA

By
V. N. FADDEEVA

Authorized translation from the Russian by
Curtis D. Benster

Dover Publications, Inc.
New York

Published in Canada by General Publishing Company, Ltd., 30 Lesmill Road, Don Mills, Toronto, Ontario.
Published in the United Kingdom by Constable and Company, Ltd., 10 Orange St., London WC 2.

Computational Methods of Linear Algebra is a new English translation from the Russian, first published by Dover Publications, Inc., in 1959.
Reproduction of this book in whole or in part by or for the United States Government is permitted for any purpose of the United States Government.

International Standard Book Number: 486-60424-1
Library of Congress Catalog Card Number: 59-8985

Manufactured in the United States of America
Dover Publications, Inc.
180 Varick Street
New York, N. Y. 10014

AUTHOR'S PREFACE

The numerical solution of the problems of mathematical physics is most frequently connected with the numerical solution of basic problems of linear algebra—that of solving a system of linear equations, and that of the computation of the proper numbers of a matrix. The present book is an endeavor at systematizing the most important numerical methods of linear algebra—classical ones and those developed quite recently as well.

The author does not pretend to an exhaustive completeness, having included an exposition of those methods only that have already been tested in practice. In the exposition the author has not strived for an irreproachable rigor, and has not analysed all conceivable cases and sub-cases arising in the application of this or that method, having limited herself to the most typical and practical important cases.

The book consists of three chapters. In the first chapter is given the material from linear algebra that is indispensable to what follows. The second chapter is devoted to the numerical solution of systems of linear equations and parallel questions. Lastly, the third chapter contains a description of numerical methods of computing the proper numbers and proper vectors of a matrix.

For the interest manifested in the manuscript, and for a number of valuable suggestions, I express my sincere thanks to A. L. Brudno and G. P. Akilov.

TRANSLATOR'S NOTE

But for the initiative of Dr. George E. Forsythe, this translation would not have been written; it is thanks to his awareness and appreciation of this work in the original, as well as to his support of the translating, that this English-language version now appears. And to Mrs. Faddeeva, the author, go my respects for a real computor's guide-book, with a nice sense of both theory and practice, and a presentation no less nice—a book such as I wish I had had.

Grateful acknowledgments are due to my father, L. Halsey Benster, for contributions beyond a mechanical engineer's line of duty (including proofreading in particular), and to my wife Ada, for more than can be expressed. June Wolfenbarger's fastidious typing helped get the manuscript off the ground.

I have added a few notes [those in brackets], replaced many of the Russian references with more accessible ones, and re-computed all of the principal tables, which I hope will thus be especially reliable and useful guides.

Curtis D. Benster

Ophir, Colorado
1958

CONTENTS

Contents

TITLES OF TABLES

Chapter II

Titles of Tables

Chapter III

COMPUTATIONAL
METHODS OF
LINEAR ALGEBRA

CHAPTER I

BASIC MATERIAL FROM LINEAR ALGEBRA

This chapter will be of an auxiliary character. Without detailed proofs, it will impart material from linear algebra that will be indispensable to an understanding of the following chapters.

§ 1. MATRICES

1. An aggregate of numbers—which are, generally speaking, complex—arranged in the form of a rectangular table, is called a *rectangular matrix*. This array will have m rows and n columns, and may be set forth in the form:

$$(1) \qquad A = \begin{bmatrix} a_{11} & a_{12} & \cdots & a_{1n} \\ a_{21} & a_{22} & \cdots & a_{2n} \\ \cdot & \cdot & \cdots & \cdot \\ a_{m1} & a_{m2} & \cdots & a_{mn} \end{bmatrix},$$

the first subscript, then, designating the row, the second designating the column, in which the *element* in point is located.

This may be abbreviated to the form:

$$A = [a_{ij}] \quad (i = 1, 2, \ldots, m; j = 1, 2, \ldots, n).$$

Two matrices are *equal* if their corresponding elements are equal.

Matrices composed of a single row are called simply *rows* (or, as they will be approached later, *row vectors*). Matrices composed of a single column are called *columns* (or *column vectors*).

If the number of rows of a matrix equals the number, n, of columns, it is called *square*, and of the nth *order*.

Among square matrices, an important role is played by *diagonal* matrices, i.e., matrices of which only the elements along the *principal (leading) diagonal* are different from zero:

$$(2) \qquad \begin{bmatrix} \alpha_1 & 0 & \dots & 0 \\ 0 & \alpha_2 & \dots & 0 \\ \cdot & \cdot & \cdot & \cdot \\ 0 & 0 & \dots & \alpha_n \end{bmatrix} = \ulcorner \alpha_1 \quad \alpha_2 \quad \dots \quad \alpha_n \lrcorner.$$

If all the numbers α_i of such a matrix are equal to each other, the matrix is said to be *scalar*:

$$(3) \qquad \begin{bmatrix} \alpha & 0 & \dots & 0 \\ 0 & \alpha & \dots & 0 \\ 0 & 0 & \dots & \alpha \end{bmatrix} = \ulcorner \alpha \lrcorner$$

and, if $\alpha = 1$, the matrix is said to be the *unit* matrix:

$$(4) \qquad \begin{bmatrix} 1 & 0 & \dots & 0 \\ 0 & 1 & \dots & 0 \\ \cdot & \cdot & \cdot & \cdot \\ 0 & 0 & \dots & 1 \end{bmatrix} = I.$$

Lastly, a matrix all of whose elements are equal to zero is called a *null* matrix, or *zero* matrix. We shall designate it by the symbol 0.

The determinant whose elements are the elements of a square matrix (without disarrangement) is said to be the *determinant of that matrix*, and we write the determinant of the matrix A as $|A|$, or often as $d(A)$.

2. *Multiplication of a matrix by a number. The addition of matrices.* A matrix whose elements are obtained by multiplying all the ele-

ments of the matrix A by a number α is called the *product* of the number α and the matrix A:

$$(5) \qquad \alpha A = \begin{bmatrix} \alpha a_{11} & \alpha a_{12} & \ldots & \alpha a_{1n} \\ \alpha a_{21} & \alpha a_{22} & \ldots & \alpha a_{2n} \\ \cdot & \cdot & \cdot & \cdot \\ \alpha a_{m1} & \alpha a_{m2} & \ldots & \alpha a_{mn} \end{bmatrix}.$$

A matrix C whose elements are the sums of the corresponding elements of A and B, matrices having like numbers of rows and columns, is called the *sum* of A and B:

$$(6) \quad A+B = \begin{bmatrix} a_{11}+b_{11} & a_{12}+b_{12} & \ldots & a_{1n}+b_{1n} \\ a_{21}+b_{21} & a_{22}+b_{22} & \ldots & a_{2n}+b_{2n} \\ \cdot & \cdot & \cdot & \cdot \\ a_{m1}+b_{m1} & a_{m2}+b_{m2} & \ldots & a_{mn}+b_{mn} \end{bmatrix}.$$

The operations introduced above have the following properties, as will be readily seen:

1. $A+(B+C) = (A+B)+C$.
2. $A+B \quad = B+A$.
3. $A+0 \quad = A$.
4. $(\alpha+\beta)A \quad = \alpha A+\beta A$.
5. $\alpha(A+B) \quad = \alpha A+\alpha B$.

Here A, B, and C are matrices; α and β are numbers—generally speaking, complex.

3. *The multiplication of matrices.* *Multiplication* of the matrices A and B is defined only on the assumption that the number of columns of matrix A equals the number of rows of matrix B. On this assumption, the elements of the *product*, $C=AB$, are defined in the following manner: the element in the ith row and the jth column

of the matrix C is equal to the sum of the products of the elements of the ith row of the matrix A by the corresponding elements of the jth column of matrix B.[1] Thus:

$$(7) \quad AB = \begin{bmatrix} a_{11} & a_{12} & \cdots & a_{1n} \\ a_{21} & a_{22} & \cdots & a_{2n} \\ \cdot & \cdot & \cdot & \cdot \\ a_{m1} & a_{m2} & \cdots & a_{mn} \end{bmatrix} \begin{bmatrix} b_{11} & b_{12} & \cdots & b_{1p} \\ b_{21} & b_{22} & \cdots & b_{2p} \\ \cdot & \cdot & \cdot & \cdot \\ b_{n1} & b_{n2} & \cdots & b_{np} \end{bmatrix}$$

$$= \begin{bmatrix} c_{11} & c_{12} & \cdots & c_{1p} \\ c_{21} & c_{22} & \cdots & c_{2p} \\ \cdot & \cdot & \cdot & \cdot \\ c_{m1} & c_{m2} & \cdots & c_{mp} \end{bmatrix} = C,$$

where

$$c_{ij} = a_{i1}b_{1j} + a_{i2}b_{2j} + \cdots + a_{in}b_{nj} = \sum_{k=1}^{n} a_{ik}b_{kj}$$

(8)

$$(i = 1, 2, \ldots, m; j = 1, 2, \ldots, p).$$

It is to be noted that the product of two rectangular matrices is again a rectangular matrix, the number of rows of which is equal to the number of rows of the first matrix, and the number of columns of which is equal to the number of columns of the second matrix: $\overset{m \times n}{A} \cdot \overset{n \times p}{B} = \overset{m \times p}{C}$. So, for instance, the product of a square matrix and a matrix composed of one column is a matrix of one column.

The commutative law for multiplication does not, generally speaking, hold. We shall make a few observations on this subject, however. The matrices AB and BA make sense simultaneously only if the number of rows of the first matrix is equal to the number of columns of the second, and the number of columns of the first is equal to the number of rows of the second. Given the fulfillment of these conditions, the matrices AB and BA will both be square, but of different orders, unless A and B be square. Thus even to put the

---

[1] [As to the motive for this apparently arbitrary definition, see, e.g., AITKEN, A. C., [2], § 3. Abundant numerical illustrations will be found in FRAZER, R. A., DUNCAN, W. J., AND COLLAR, A. R.,[1], § 1.4.]

question of the equality of the matrices AB and BA makes sense only for square matrices. But even in this case, generally speaking, $AB \neq BA$.

In particular cases multiplication may be commutative, and in such cases the matrices are said to *commute*. Thus, for example, scalar matrices commute with any square matrix of the same order, for

$$
(9) \quad
\begin{bmatrix}
\alpha & 0 & \ldots & 0 \\
0 & \alpha & \ldots & 0 \\
\cdot & \cdot & \cdot & \cdot \\
0 & 0 & \ldots & \alpha
\end{bmatrix}
\begin{bmatrix}
a_{11} & a_{12} & \ldots & a_{1n} \\
a_{21} & a_{22} & \ldots & a_{2n} \\
\cdot & \cdot & \cdot & \cdot \\
a_{n1} & a_{n2} & \ldots & a_{nn}
\end{bmatrix}
$$

$$
=
\begin{bmatrix}
a_{11} & a_{12} & \ldots & a_{1n} \\
a_{21} & a_{22} & \ldots & a_{2n} \\
\cdot & \cdot & \cdot & \cdot \\
a_{n1} & a_{n2} & \ldots & a_{nn}
\end{bmatrix}
\begin{bmatrix}
\alpha & 0 & \ldots & 0 \\
0 & \alpha & \ldots & 0 \\
\cdot & \cdot & \cdot & \cdot \\
0 & 0 & \ldots & \alpha
\end{bmatrix}
$$

$$
=
\begin{bmatrix}
\alpha a_{11} & \alpha a_{12} & \ldots & \alpha a_{1n} \\
\alpha a_{21} & \alpha a_{22} & \ldots & \alpha a_{2n} \\
\cdot & \cdot & \cdot & \cdot \\
\alpha a_{n1} & \alpha a_{n2} & \ldots & \alpha a_{nn}
\end{bmatrix}.
$$

Hence follows the special role of the unit matrix in the multiplication of matrices, to wit: amongst all square matrices of the same order, the unit matrix plays the same role as the number one does among numbers. Indeed,

$$AI = IA = A.$$

It can be shown that the multiplication of matrices is *associative*, viz.: if AB and $(AB)C$ make sense, so also do BC and $A(BC)$, and

$$1. \quad A(BC) = (AB)C.$$

The matrix product has also these properties:

$$2. \quad \alpha(AB) = (\alpha A)B = A(\alpha B);$$

$$3. \quad (A+B)C = AC+BC;$$

$$4. \quad C(A+B) = CA+CB,$$

where A, B, C are matrices, α a number.

Let us interchange rows and columns in the matrix

$$A = \begin{bmatrix} a_{11} & a_{12} & \cdots & a_{1n} \\ a_{21} & a_{22} & \cdots & a_{2n} \\ \cdot & \cdot & \cdots & \cdot \\ a_{m1} & a_{m2} & \cdots & a_{mn} \end{bmatrix} = [a_{ij}];$$

we obtain the *transposed* matrix or *transpose*

$$(10) \quad A' = A^T = [a_{ij}]' = \begin{bmatrix} a_{11} & a_{21} & \cdots & a_{m1} \\ a_{12} & a_{22} & \cdots & a_{m2} \\ \cdot & \cdot & \cdots & \cdot \\ a_{1n} & a_{2n} & \cdots & a_{mn} \end{bmatrix} = [a_{ji}].$$

The following rule (the *reversal rule*) for a transposed product should be noted:

$$(AB)' = B'A'$$

In proof of this, note that the element of the ith row and jth column of the matrix $(AB)'$ is equal to the element of the jth row and ith column of the matrix AB, for this is merely the interchange of row with column, i.e., transposition; and that is equal to

$$(11) \qquad a_{j1}b_{1i}+a_{j2}b_{2i}+ \cdots +a_{jn}b_{ni}.$$

The last expression is obviously equal to the sum of the products of the elements of the ith row of the matrix B' and the corresponding elements of the jth column of the matrix A', i.e., is equal to the general element of the matrix $B'A'$.

In conclusion we shall remark that the determinant of a product of square matrices is equal to the product of the determinants of the multiplied matrices: $|AB| = |A|\,|B|$, which result is taken from determinant theory.

4. *The partitioning of matrices.* The handling of matrices of high orders requires, as a rule, a large number of operations. It is therefore often convenient to reduce a computation involving matrices of high orders to computations upon matrices of lower orders. Such a reduction can be effected by *partitioning* the given matrices: each matrix may be conceived as composed of several matrices of lower orders, and this subdivision may be carried through in many ways, for example:

$$
\begin{bmatrix}
a_{11} & a_{12} & a_{13} & a_{14} \\
a_{21} & a_{22} & a_{23} & a_{24} \\
a_{31} & a_{32} & a_{33} & a_{34}
\end{bmatrix}
=
\left[\begin{array}{c:ccc}
a_{11} & a_{12} & a_{13} & a_{14} \\ \hdashline
a_{21} & a_{22} & a_{23} & a_{24} \\
a_{31} & a_{32} & a_{33} & a_{34}
\end{array}\right]
$$

$$
=
\left[\begin{array}{cc:cc}
a_{11} & a_{12} & a_{13} & a_{14} \\ \hdashline
a_{21} & a_{22} & a_{23} & a_{24} \\
a_{31} & a_{32} & a_{33} & a_{34}
\end{array}\right]
$$

The matrices into which the given matrix is partitioned are called its *submatrices*, or *cells*. In such a partition the horizontal and vertical subdividing lines are of course supposed to be carried across the whole matrix.

We shall not concern ourselves with the general case of the partitioning of a matrix,[1] but shall here consider only a partition of square matrices in which the diagonal submatrices are square.

The basic operations on partitioned matrices whose diagonal matrices are of identical orders is connected in a quite natural way

[1] [See, e.g., AITKEN, A. C., [2], § 11.]

with operations upon the submatrices themselves. To wit, if we have

$$A = \begin{bmatrix} A_{11} & A_{12} & \ldots & A_{1k} \\ A_{21} & A_{22} & \ldots & A_{2k} \\ \cdot & \cdot & \cdot & \cdot \cdot \cdot \cdot \cdot \\ A_{k1} & A_{k2} & \ldots & A_{kk} \end{bmatrix}$$

and

$$B = \begin{bmatrix} B_{11} & B_{12} & \ldots & B_{1k} \\ B_{21} & B_{22} & \ldots & B_{2k} \\ \cdot & \cdot & \cdot & \cdot \cdot \cdot \cdot \cdot \\ B_{k1} & B_{k2} & \ldots & B_{kk} \end{bmatrix},$$

where A_{ii} and B_{ii} are square matrices of the same order, then

$$(12) \quad A+B = \begin{bmatrix} A_{11}+B_{11} & A_{12}+B_{12} & \ldots & A_{1k}+B_{1k} \\ A_{21}+B_{21} & A_{22}+B_{22} & \ldots & A_{2k}+B_{2k} \\ \cdot & \cdot & \cdot & \cdot \cdot \cdot \cdot \cdot \\ A_{k1}+B_{k1} & A_{k2}+B_{k2} & \ldots & A_{kk}+B_{kk} \end{bmatrix}$$

and

$$(13) \quad AB = \begin{bmatrix} C_{11} & C_{12} & \ldots & C_{1k} \\ C_{21} & C_{22} & \ldots & C_{2k} \\ \cdot & \cdot & \cdot & \cdot \cdot \cdot \\ C_{k1} & C_{k2} & \ldots & C_{kk} \end{bmatrix},$$

where

$$C_{ij} = A_{i1}B_{1j}+A_{i2}B_{2j}+ \cdots +A_{ik}B_{kj} \quad (i,j=1,\ldots,k).$$

We shall not stop for a proof of the last formula, but will note here only that the matrices A_{i1} and B_{1j} can indeed be multiplied, since the number of columns of matrix A_{i1} equals the number of rows of the matrix B_{1j}.

Formulas (12) and (13) show that operations with matrices partitioned in the manner indicated are to be performed just as if in place of each submatrix there was a number.

An important special case of a partitioned matrix is the *bordered* matrix. Having a square matrix A_{n-1} of order $n-1$:

$$A_{n-1} = \begin{bmatrix} a_{11} & a_{12} & \cdots & a_{1,\,n-1} \\ a_{21} & a_{22} & \cdots & a_{2,\,n-1} \\ \cdot & \cdot \cdot \cdot \cdot \cdot \cdot & \cdot \cdot \cdot & \cdot \cdot \\ a_{n-1,\,1} & a_{n-1,\,2} & \cdots & a_{n-1,\,n-1} \end{bmatrix},$$

we form a square matrix of the nth order, A_n, by appending to the matrix A_{n-1} a row: $v_{n-1} = (a_{n1}, \ldots, a_{n,\,n-1})$, a column: $u_{n-1} = (a_{1n}, \ldots, a_{n-1,\,n})$, and a number a_{nn}:

$$(14) \quad A_n = \begin{bmatrix} & & & a_{1n} \\ A_{n-1} & & & a_{2n} \\ & & & \vdots \\ & & & a_{n-1,\,n} \\ a_{n1} & \cdots & a_{n,\,n-1} & a_{nn} \end{bmatrix} = \begin{bmatrix} A_{n-1} & u_{n-1} \\ v_{n-1} & a_{nn} \end{bmatrix}.$$

We shall say that the matrix A_n has been obtained by *bordering* the matrix A_{n-1}. The matrix A_n is naturally partitionable.

Operations upon a bordered matrix are performed in accordance with the general rules for operations upon partitioned matrices. Letting

$$A = \begin{bmatrix} M & u \\ v & a \end{bmatrix}, \qquad B = \begin{bmatrix} P & y \\ x & b \end{bmatrix}$$

be two bordered matrices of order n, the meaning of M, v, u, a, and P, x, y, b, being those of the definition, the following statements are valid:

$$(15) \quad \begin{cases} \alpha A = \begin{bmatrix} \alpha M & \alpha u \\ \alpha v & \alpha a \end{bmatrix}, \\[2ex] A + B = \begin{bmatrix} M+P & u+y \\ v+x & a+b \end{bmatrix}, \\[2ex] AB = \begin{bmatrix} MP+ux & My+ub \\ vP+ax & vy+ab \end{bmatrix}. \end{cases}$$

Here MP and ux are matrices of the $(n-1)$th order; My, ub are columns composed of $n-1$ elements; vP and ax are analogous rows; and, lastly, $vy + ab$ is a number.

5. *Quasi-diagonal matrices.* Let us consider still another particular case of partitioned matrices, namely the matrices called quasi-diagonal. These are square matrices along whose leading diagonals are arrayed square submatrices, the remainder of the elements being zero. An example would be the seventh-order quasi-diagonal matrix:

$$\begin{bmatrix} a_{11} & a_{12} & 0 & 0 & 0 & 0 & 0 \\ a_{21} & a_{22} & 0 & 0 & 0 & 0 & 0 \\ 0 & 0 & b_{11} & b_{12} & b_{13} & 0 & 0 \\ 0 & 0 & b_{21} & b_{22} & b_{23} & 0 & 0 \\ 0 & 0 & b_{31} & b_{32} & b_{33} & 0 & 0 \\ 0 & 0 & 0 & 0 & 0 & c_{11} & c_{12} \\ 0 & 0 & 0 & 0 & 0 & c_{21} & c_{22} \end{bmatrix}$$

The cells of this matrix are obviously

$$A = \begin{bmatrix} a_{11} & a_{12} \\ a_{21} & a_{22} \end{bmatrix};$$

$$B = \begin{bmatrix} b_{11} & b_{12} & b_{13} \\ b_{21} & b_{22} & b_{23} \\ b_{31} & b_{32} & b_{33} \end{bmatrix}; \qquad C = \begin{bmatrix} c_{11} & c_{12} \\ c_{21} & c_{22} \end{bmatrix}$$

and the six null submatrices.

If two quasi-diagonal matrices are of like structure, the product of such matrices will also be a quasi-diagonal matrix of the same structure, the diagonal cells of which equal the products of the corresponding cells of the factor matrices.

The determinant of a quasi-diagonal matrix is equal to the product of the determinants of the diagonal cells, on the strength of a notable theorem by Laplace.[1]

[1] [See, e.g., AITKEN, A. C., [2] , § 33.]

6. *The inverse and the adjoint matrices.* A square matrix $A = [a_{ij}]$ is said to be *non-singular* if its determinant is not equal to zero; in the contrary case it is of course *singular*.

The important concept of an inverse matrix is now introduced. A matrix B is called the *inverse* (or *reciprocal*) of the matrix A if

$$(16) \qquad AB = I.$$

We shall show that the necessary and sufficient condition for the existence of the inverse matrix is the non-singularity of the matrix A.

The necessity follows at once from the theorem concerning the determinant of a matrix product, for if $AB = I$, $|A|\,|B| = 1$ and consequently $|A| \neq 0$.

Assume now that $|A| \neq 0$. In order to construct the inverse matrix we must give preliminary consideration to the *adjoint* (or *adjugate*) matrix, i.e., the matrix

$$(17) \qquad C = \begin{bmatrix} A_{11} & A_{21} & \dots & A_{n1} \\ A_{12} & A_{22} & \dots & A_{n2} \\ \cdot & \cdot & \cdot & \cdot \\ A_{1n} & A_{2n} & \dots & A_{nn} \end{bmatrix}$$

Here A_{ij} is the algebraic complement (cofactor) of the element a_{ij} in the determinant of the matrix A, i.e., is the signed minor determinant of the element a_{ij}. Note that it is here placed in transposed position.

We shall show that the adjoint matrix has the following property:

$$(18) \qquad AC = |A|I$$

In demonstration, reckoning the general element of the matrix AC by the rule for matrix multiplication, we find it to equal

$$a_{i1}A_{j1} + a_{i2}A_{j2} + \cdots + a_{in}A_{jn},$$

i.e., zero for $i \neq j$, and $|A|$ for $i = j$, on the strength of a familiar theorem on the expansion of determinants.[1]

[1] [See, e.g., AITKEN, A. C., [2], § 21.]

The equality

(18')
$$CA = |A|I$$

is established in like manner.

The adjoint matrix has meaning for any square matrix A. From the equality $AC = |A|I$ it follows that the matrix

(19)
$$B = \frac{1}{|A|}C$$

is, for non-singular A, the sought inverse, for

$$AB = A\frac{1}{|A|}C = \frac{1}{|A|}AC = I.$$

The constructed matrix has also the property

(20)
$$BA = I,$$

which follows from equation (18').

We prove, lastly, the uniqueness of the inverse matrix. Assume that a matrix X exists such that $AX = I$. Multiplying this equation by B on the left, we have $X = B$. If it be assumed that $YA = I$, a multiplication on the right by B yields $Y = B$.

The matrix inverse to A is denoted by A^{-1}. It is obvious that $|A^{-1}| = |A|^{-1}$.

We note that the inverse of the product of two matrices also displays the reversal rule:

(21)
$$(A_1A_2)^{-1} = A_2^{-1}A_1^{-1},$$

since

$$A_1A_2A_2^{-1}A_1^{-1} = A_1A_1^{-1} = I.$$

The determination of the inverse matrix is one of the fundamental problems of linear algebra. Equation (19) offers the possibility of computing the inverse matrix; however, the computation of the adjoint matrix is so labor-consuming that the cited equation is of importance only in theoretical relationships. Chapter II will be specially devoted to this problem of determining the inverse matrix.

7. *Polynomials in a matrix.* We now define the positive integral power of a square matrix, putting

$$(22) \qquad \overbrace{A \cdot A \cdot \ \ldots \cdot A}^{n \text{ times}} = A^n.$$

In view of the associative law, how the parentheses in this product are placed makes no difference, and we therefore omit them. It is evident from the definition that

$$(23) \qquad \begin{cases} A^n A^m = A^{n+m} \\ (A^n)^m = A^{nm}. \end{cases}$$

Hence it follows that *powers of the same matrix are commutative.*

We further put, by definition,

$$A^0 = I.$$

An expression of the form

$$\alpha_0 A^n + \alpha_1 A^{n-1} + \cdots + \alpha_n I,$$

where $\alpha_0, \alpha_1, \ldots, \alpha_n$ are complex numbers, is called a *polynomial in a matrix*, or *matrix polynomial*. This matrix polynomial may be regarded as the result of replacing the variable λ in an algebraic polynomial

$$(24) \qquad \varphi(\lambda) = \alpha_0 \lambda^n + \alpha_1 \lambda^{n-1} + \cdots + \alpha_n$$

by the matrix A.

It is important to note that the rules for operation upon matrix polynomials do not differ from the rules for operation upon algebraic polynomials, viz.:

$$(25) \qquad \begin{cases} \text{given} & \varphi(\lambda) = \psi(\lambda) \pm \chi(\lambda) \\ & \omega(\lambda) = \psi(\lambda)\, \chi(\lambda), \\ \text{then} & \varphi(A) = \psi(A) \pm \chi(A) \\ & \omega(A) = \psi(A)\, \chi(A). \end{cases}$$

This follows from the commutativity of the powers of a matrix.

8. *The characteristic polynomial.   The Cayley-Hamilton theorem.   The minimum polynomial.*   The equation

(26)
$$\begin{vmatrix} a_{11}-\lambda & a_{12} & \cdots & a_{1n} \\ a_{21} & a_{22}-\lambda & \cdots & a_{2n} \\ \cdot\cdot\cdot\cdot\cdot\cdot\cdot\cdot\cdot\cdot\cdot\cdot\cdot \\ a_{n1} & a_{n2} & \cdots & a_{nn}-\lambda \end{vmatrix} = 0$$

is called the *characteristic*, or *secular*, *equation* of the matrix $A = (a_{ij})$. The left member of this equation, which may be written in the abbreviated form $|A - \lambda I|$, bears the name *characteristic polynomial* (or *characteristic function*) of the matrix.   Characteristic equations are frequently encountered in applied mathematics.

The direct computation of the characteristic function presents considerable technical difficulties.   If

(27) $\varphi(\lambda) = |A - \lambda I| = (-1)^n[\lambda^n - p_1\lambda^{n-1} - p_2\lambda^{n-2} - \cdots - p_n],$

then

$$p_1 = a_{11} + a_{22} + \cdots + a_{nn},$$

(28)

$$p_n = (-1)^{n-1}|A|,$$

and the remaining coefficients p_k are the sums, taken with the sign $(-1)^{k-1}$, of all the *principal minors* of the determinant of matrix A of order k, i.e., of the minors involving the principal diagonal.[1]   The number of such minors equals the number of combinations of n things taken k at a time.

The roots of the characteristic equation are called the *proper numbers* (or *characteristic roots, latent roots, proper values, eigenvalues*) of the matrix A.   From the well-known theorem of Vietà giving the connection between the roots of an equation and its coefficients, we have

(29) $\begin{cases} \lambda_1 + \lambda_2 + \cdots + \lambda_n = p_1 = a_{11} + a_{22} + \cdots + a_{nn} \\ \lambda_1\lambda_2 \ldots \lambda_n = (-1)^{n-1}p_n = |A|. \end{cases}$

[1] [See, e.g., AITKEN, A. C., [2], § 37.]

The quantity $p_1 = a_{11} + a_{22} + \cdots + a_{nn}$ is called the *trace* (or *spur*) of the matrix A, and is denoted by tr A.

Practically convenient methods for determining the coefficients and roots of the characteristic equation will be elaborated in Chapter III, which will be specially devoted to that group of questions. For the moment we leave them aside.

For any square matrix the following remarkable relation, known as the *Cayley-Hamilton theorem*, obtains: if $\varphi(\lambda)$ is the characteristic polynomial of the matrix A, then $\varphi(A) = 0$, that is, in a sense, *the matrix is a root of its own characteristic equation.*

For proof, let us consider the matrix B, the adjoint of the matrix $A - \lambda I$. Since each cofactor in the determinant $|A - \lambda I|$ is a polynomial in λ of degree not exceeding $n - 1$, the adjoint matrix may be represented as an algebraic polynomial with matrix coefficients, i.e., in the form

$$B = B_{n-1} + B_{n-2}\lambda + \cdots + B_0\lambda^{n-1},$$

where $B_{n-1}, \ldots, B_0$ are certain matrices not dependent on λ. On the strength of the fundamental property of the adjoint matrix, we have

$$(B_{n-1} + B_{n-2}\lambda + \cdots + B_0\lambda^{n-1})(A - \lambda I) = |A - \lambda I|I$$
$$= (-1)^n(\lambda^n - p_1\lambda^{n-1} - \cdots - p_n)I.$$

This equation is equivalent to the system of equations

$$\begin{cases} B_{n-1}A & = (-1)^{n+1}p_nI \\ B_{n-2}A - B_{n-1} & = (-1)^{n+1}p_{n-1}I \\ \cdots\cdots\cdots\cdots\cdots\cdots \\ B_0A - B_1 & = (-1)^{n+1}p_1I \\ -B_0 & = (-1)^nI \end{cases}$$

Multiplying these equations on the right by $I, A, A^2, \ldots, A^{n-1}, A^n$, respectively, and adding, we obtain a null matrix on the left side, and on the right,

$$(30) \qquad (-1)^n[-p_nI - p_{n-1}A - p_{n-2}A^2 - \cdots + A^n] = \varphi(A).$$

Thus $\varphi(A) = 0$, which is what was required to be proved.

The Cayley-Hamilton relation shows that for a given square matrix a polynomial exists for which it is a root. Evidently such a polynomial is not unique, for if $\psi(\lambda)$ has such a property, so has any polynomial divisible by $\psi(\lambda)$. The polynomial of lowest degree having the property that the matrix A is a zero of it, is called the *minimum polynomial* of the matrix A.

We shall prove that the characteristic polynomial is divisible by the minimum polynomial.

Let $q(\lambda)$ and $r(\lambda)$ be the quotient and remainder obtained upon dividing the characteristic polynomial $\varphi(\lambda)$ by the minimum polynomial $\psi(\lambda)$:

$$\varphi(\lambda) = \psi(\lambda)\, q(\lambda) + r(\lambda),$$

the degree of $r(\lambda)$ being of course less than the degree of $\psi(\lambda)$.

Substituting A for λ in this equation, we have

$$r(A) = \varphi(A) - \psi(A)\, q(A) = 0.$$

Thus the matrix A proves to be a "zero" of the polynomial $r(\lambda)$; it thence follows that $r(\lambda) \equiv 0$, since otherwise $\psi(\lambda)$ would not be the minimum polynomial. Consequently $\psi(\lambda)$ divides $\varphi(\lambda)$.

9. *Similar matrices.* The matrix B is said to be *similar* to the matrix A if a non-singular matrix C exists such that $B = C^{-1}AC$. Matrix B is said to be obtained from matrix A by a *similarity transformation*.

The similarity transformation has the following properties:

$$\text{1.} \quad C^{-1}A_1C + C^{-1}A_2C + \cdots + C^{-1}A_nC$$
$$= C^{-1}(A_1 + A_2 + \cdots + A_n)C.$$

$$(31) \quad \text{2.} \quad C^{-1}A_1C \cdot C^{-1}A_2C \cdot \ldots \cdot C^{-1}A_nC = C^{-1}(A_1A_2\ldots A_n)C.$$

In particular, $(C^{-1}AC)^n = C^{-1}A^nC$.
Hence:

$$\text{3.} \quad f(C^{-1}AC) = C^{-1}f(A)\,C \quad \text{for any polynomial } f(\lambda).$$

From the last property it follows directly that *similar matrices have the same minimum function.*

We shall show that *similar matrices have also the same characteristic function.*

We have

$$|B - \lambda I| = |C^{-1}AC - \lambda I| = |C^{-1}AC - \lambda C^{-1}IC|$$

$$= |C|^{-1}|A - \lambda I| |C| = |A - \lambda I|.$$

10. *Elementary transformations.* It is frequently necessary to effect the following operations upon matrices:

a) Multiplication of the elements of some row by a number;

b′) Adding to the elements of some row numbers proportional to the elements of some preceding row.

b″) Adding to the elements of some row numbers proportional to the elements of some following row.

Sometimes such transformations must be made upon the columns. Transformations of the type indicated are called *elementary transformations* of the matrix.

Any elementary transformation of the *rows* is equivalent to a *pre*multiplication of the matrix (multiplication on the left) by a nonsingular matrix of a special form, as will readily be verified. Operation **a**) is equivalent to a premultiplication of the matrix by the matrix

$$(32) \qquad \begin{bmatrix} 1 & & & & & & \\ & \cdot & & & & & \\ & & \cdot & & & & \\ & & & \cdot & & & \\ & & & & 1 & & \\ & & & & \alpha & & \\ & & & & & 1 & \\ & & & & & & \cdot \\ & & & & & & & \cdot \\ & & & & & & & & \cdot \\ & & & & & & & & & 1 \end{bmatrix} ;$$

operation **b′**) is equivalent to a premultiplication by the matrix

$$(33) \qquad \begin{bmatrix} 1 & & & & & & & \\ & 1 & & & & & & \\ & & \ddots & & & & & \\ & & & \ddots & & & & \\ & & & & \ddots & & & \\ \alpha & \cdots & \cdots & \cdots & \cdots & 1 & & \\ & & & & & & \ddots & \\ & & & & & & & \ddots \\ & & & & & & & & 1 \end{bmatrix} ;$$

operation **b''**) is equivalent to a premultiplication by the matrix

$$(34) \qquad \begin{bmatrix} 1 & & & & & & \\ & \ddots & & & & & \\ & & 1 & \cdots & \cdots & \alpha & \\ & & & \ddots & & & \\ & & & & \ddots & & \\ & & & & & 1 & \\ & & & & & & \ddots \\ & & & & & & & 1 \end{bmatrix}$$

Operations **a**), **b'**) and **b''**) are performed on columns by using just such *elementary matrices*[1] in *post*multiplications of the matrix under-

---

[1] [Note that the elementary matrices may well be regarded as having been derived from the unit matrix by just such transformations, **a**), **b'**), and **b''**), as it is proposed to make upon the rows/columns of A. Thus, for example, the second elementary matrix of the example effects, by premultiplication, the following transformation of A: row $3 + \alpha$ row 2, and is itself the result of such an operation upon the unit matrix. By postmultiplication it effects: column $2 + \alpha$ column 3, and is this transformation of I.]

going transformation, (33) now effecting **b″**) upon columns, and (34) effecting **b′**).

Examples of row operations:

$$
\begin{bmatrix} 1 & 0 & 0 \\ 0 & \alpha & 0 \\ 0 & 0 & 1 \end{bmatrix}
\begin{bmatrix} a & b & c \\ x & y & z \\ u & v & w \end{bmatrix}
=
\begin{bmatrix} a & b & c \\ \alpha x & \alpha y & \alpha z \\ u & v & w \end{bmatrix};
$$

$$
\begin{bmatrix} 1 & 0 & 0 \\ 0 & 1 & 0 \\ 0 & \alpha & 1 \end{bmatrix}
\begin{bmatrix} a & b & c \\ x & y & z \\ u & v & w \end{bmatrix}
=
\begin{bmatrix} a & b & c \\ x & y & z \\ u+\alpha x & v+\alpha y & w+\alpha z \end{bmatrix};
$$

$$
\begin{bmatrix} 1 & 0 & 0 \\ 0 & 1 & \alpha \\ 0 & 0 & 1 \end{bmatrix}
\begin{bmatrix} a & b & c \\ x & y & z \\ u & v & w \end{bmatrix}
=
\begin{bmatrix} a & b & c \\ x+\alpha u & y+\alpha v & z+\alpha w \\ u & v & w \end{bmatrix}.
$$

In future work we will often have to perform transformations of types **a**) and **b′**) upon matrices. The result of several row-transformations of this type is equivalent to the premultiplication of the matrix by some *triangular* matrix, i.e., one of the form

$$
(35) \qquad
\begin{bmatrix}
\gamma_{11} & 0 & \cdots & 0 \\
\gamma_{21} & \gamma_{22} & \cdots & 0 \\
\cdot & \cdot & \cdots & \cdot \\
\gamma_{n1} & \gamma_{n2} & \cdots & \gamma_{nn}
\end{bmatrix},
$$

with non-zero diagonal elements γ_{ii}. Indeed, each separate transformation of form **a**) or **b′**) is equivalent to a premultiplication by a triangular matrix of the type indicated, and the product of two or more triangular matrices of like structure (i.e., both, say, *lower triangular*, as these) is again a like triangular matrix.

It should be further noted that the result of several column-transformations of the form **b′**) and **b″**), such that to each, *j*th, column is added a multiple, m_{ij}, of the elements of the *i*th column (which itself remains unchanged), is equivalent to postmultiplication of the given matrix by a matrix of the form:

$$(36) \quad \begin{bmatrix} 1 & 0 & . & . & . & . & . & . & . & . & . & . & . & . & . & 0 \\ 0 & 1 & . & . & . & . & . & . & . & . & . & . & . & . & . & 0 \\ . & . & . & . & . & . & . & . & . & . & . & . & . & . & . & . \\ m_{i1} & m_{i2} & . . . & m_{i,\,i-1} & & 1 & & m_{i,\,i+1} & . . . & m_{in} \\ . & . & . & . & . & . & . & . & . & . & . & . & . & . & . & . \\ 0 & 0 & . & . & . & . & . & . & . & . & . & . & . & . & . & 1 \end{bmatrix} .$$

11. *Decomposition of matrices into the product of two triangular matrices.* Triangular matrices, that is, matrices of the form

$$(37) \quad \begin{bmatrix} c_{11} & 0 & . . . & 0 \\ c_{21} & c_{22} & . . . & 0 \\ . & . & . & . & . & . & . \\ c_{n1} & c_{n2} & . . . & c_{nn} \end{bmatrix} \quad \text{and} \quad \begin{bmatrix} b_{11} & b_{12} & . . . & b_{1n} \\ 0 & b_{22} & . . . & b_{2n} \\ . & . & . & . & . & . & . \\ 0 & 0 & . . . & b_{nn} \end{bmatrix} ,$$

$$\text{(no } c_{ii} = 0, \text{ no } b_{ii} = 0)$$

have a number of convenient properties. For instance, the determinant of a triangular matrix equals the product of the elements of the principal diagonal; the product of two triangular matrices of like structure is again a triangular matrix of the same structure; a non-singular triangular matrix is easily inverted and its inverse is of like structure, etc.

The following theorems are therefore of interest.

THEOREM. *On condition that the leading submatrices of the matrix*

$$A = \begin{bmatrix} a_{11} & a_{12} & . . . & a_{1n} \\ a_{21} & a_{22} & . . . & a_{2n} \\ . & . & . & . & . & . & . \\ a_{n1} & a_{n2} & . . . & a_{nn} \end{bmatrix}$$

are non-singular, i.e., that

$$a_{11} \neq 0, \quad \begin{vmatrix} a_{11} & a_{12} \\ a_{21} & a_{22} \end{vmatrix} \neq 0, \ldots, \quad |A| \neq 0,$$

A may be represented as the product of a lower triangular matrix and an upper triangular matrix.

The proof will be carried through by the method of mathematical induction.

For $n = 1$, the statement is obvious: $(a_{11}) = (b_{11})(c_{11})$, and one of the factors may be taken arbitrarily. Let the theorem be true for a matrix of the $(n-1)$th order. We shall show it to be true for a matrix of the nth order.

Partition the matrix A into a bordered matrix:

$$A = \begin{bmatrix} a_{11} & \cdots & a_{1n} \\ a_{21} & \cdots & a_{2n} \\ \cdot & \cdot \cdot \cdot \cdot & \cdot \\ a_{n1} & \cdots & a_{nn} \end{bmatrix} = \begin{bmatrix} & & & a_{1n} \\ & A_{n-1} & & \vdots \\ & & & a_{n-1,\,n} \\ a_{n1} & a_{n2} & \cdots & a_{n,\,n-1} & a_{nn} \end{bmatrix}$$

$$= \begin{bmatrix} A_{n-1} & u \\ v & a_{nn} \end{bmatrix};$$

we shall seek a decomposition $A = CB$ of the matrix A into the product of two matrices B and C of the required forms, first having partitioned these matrices into bordered form like that of A:

$$C = \begin{bmatrix} C_{n-1} & 0 \\ x & c_{nn} \end{bmatrix}, \qquad B = \begin{bmatrix} B_{n-1} & y \\ 0 & b_{nn} \end{bmatrix}.$$

By the rule for multiplication of partitioned matrices:

$$CB = \begin{bmatrix} C_{n-1} & 0 \\ x & c_{nn} \end{bmatrix} \begin{bmatrix} B_{n-1} & y \\ 0 & b_{nn} \end{bmatrix}$$

$$= \begin{bmatrix} C_{n-1}B_{n-1} & C_{n-1}y \\ xB_{n-1} & xy + c_{nn}b_{nn} \end{bmatrix} = A,$$

whence we have

$$C_{n-1}B_{n-1} = A_{n-1}.$$

Now such triangular matrices, C_{n-1} and B_{n-1}, exist, by the induction hypothesis. Furthermore, from the assumption that $|A_{n-1}| \neq 0$, it follows that $|C_{n-1}| \neq 0$ and $|B_{n-1}| \neq 0$.

Now x and y are found by the formulas

$$y = C_{n-1}^{-1}u, \qquad x = vB_{n-1}^{-1},$$

wherewith they are determined uniquely in terms of u and v.

Thus it only remains for us to determine the diagonal elements c_{nn} and b_{nn} from the equation

$$c_{nn}b_{nn} = a_{nn} - xy.$$

The last equation shows that one of the diagonal elements may be taken arbitrarily.

Thus *the decomposition of a matrix into the product of two triangular matrices, lower and upper, will be unique if we prescribe values for the diagonal elements of one of the triangular matrices.*

It is convenient to consider, for example, that $b_{ii} = 1, i = 1, \ldots, n$. Then

$$c_{nn} = a_{nn} - xy$$

and accordingly matrix C will be uniquely determined.

12. *Matrix notation for a system of linear equations.* Let us consider the system of n linear equations in n unknowns:

$$(38) \quad \begin{cases} a_{11}x_1 + a_{12}x_2 + \cdots + a_{1n}x_n = b_1 \\ a_{21}x_1 + a_{22}x_2 + \cdots + a_{2n}x_n = b_2 \\ \qquad \cdots \cdots \cdots \cdots \cdots \cdots \\ a_{n1}x_1 + a_{n2}x_2 + \cdots + a_{nn}x_n = b_n. \end{cases}$$

Utilizing the definition of matrix multiplication, the system may be written as a single equation in matrix notation:

$$(38') \quad \begin{bmatrix} a_{11} & a_{12} & \cdots & a_{1n} \\ a_{21} & a_{22} & \cdots & a_{2n} \\ \cdots & \cdots & \cdots & \cdots \\ a_{n1} & a_{n2} & \cdots & a_{nn} \end{bmatrix} \begin{bmatrix} x_1 \\ x_2 \\ \vdots \\ x_n \end{bmatrix} = \begin{bmatrix} b_1 \\ b_2 \\ \vdots \\ b_n \end{bmatrix}$$

or simply as

$$(38'') \qquad\qquad Ax = b,$$

A signifying the matrix of the coefficients of the system, b the column of constant terms, and x the column whose elements are the unknowns.

If the matrix of the system, A, is non-singular, we obtain at once the solution of system (38) by premultiplying (38″) by A^{-1}:

$$(39) \qquad x = A^{-1}b = \frac{1}{|A|} Bb,$$

B being the adjoint matrix of A.

We shall show that the last formula is the matrix notation for the familiar *Cramer's Rule*:

$$(40) \qquad x_i = \frac{|A_i|}{|A|},$$

where A_i is the matrix that is obtained from A by replacing the elements a_{ki} of the ith column by the components b_i of b.

Indeed, the matrix equation (39) is equivalent to the n equations

$$(40a) \qquad x_i = \frac{A_{1i}b_1 + A_{2i}b_2 + \cdots + A_{ni}b_n}{|A|} \quad (i = 1, \ldots, n).$$

Since the A_{ki} are the cofactors of the element a_{ki} in the determinant of the matrix A, we obviously have

$$(40b) \qquad A_{1i}b_1 + A_{2i}b_2 + \cdots + A_{ni}b_n = |A_i|,$$

which proves our statement.

§ 2. *n*-DIMENSIONAL VECTOR SPACE

In what is to follow an important role will be played by the so-called *n-dimensional vector space R_n*. A *point X* of such a space is an aggregate of n numbers, as a rule complex, arrayed in a definite order:

$$(1) \qquad X = (x_1, x_2, \ldots, x_n).$$

X is also called an *n-dimensional vector*. The numbers $x_1, x_2, \ldots, x_n$ are called the *components* of the vector. The number n is called the *dimension* of the space.

Two vectors are said to be equal only if their corresponding components are equal. Fundamental operations on vectors are defined as follows: If $X = (x_1, x_2, \ldots, x_n)$ and $Y = (y_1, y_2, \ldots, y_n)$ are two n-dimensional vectors and a is an arbitrary complex number, we then put, by definition,

$$(2) \qquad \begin{cases} X + Y = (x_1 + y_1, x_2 + y_2, \ldots, x_n + y_n) \\ aX = (ax_1, ax_2, \ldots, ax_n). \end{cases}$$

The addition of vectors satisfies the commutative and associative laws:

$$X + Y = Y + X$$
$$(X + Y) + Z = X + (Y + Z).$$

The addition of vectors is connected with multiplication by numbers by the distributive laws

$$(3) \qquad \begin{aligned} a(X + Y) &= aX + aY \\ (a + b)X &= aX + bX. \end{aligned}$$

The validity of all these laws follows directly from the definition of the operations.

For vectors of an n-dimensional space a scalar product is introduced in accordance with the formula

$$(4) \qquad (X, Y) = \sum_{k=1}^{n} x_k \bar{y}_k ,$$

where $\bar{y}_k$ designates the complex conjugate of y_k.

It is readily verified that the scalar product has the following properties:

1) $(X, X) > 0$ if $X \neq 0$; $(X, X) = 0$ if $X = 0$.
2) $(X, Y) = \overline{(Y, X)}$.
3) $(X_1 + X_2, Y) = (X_1, Y) + (X_2, Y)$.
4) $(aX, Y) = a(X, Y)$.
5) $(X, Y_1 + Y_2) = (X, Y_1) + (X, Y_2)$.
6) $(X, aY) = \bar{a}(X, Y)$.

In addition, $\sqrt{(X, X)}$ is called the *length* of the vector. In what follows we shall designate it by $\|X\|$.

Besides the n-dimensional complex space introduced above, it is useful to consider also an n-dimensional real space, i.e., the aggregate of vectors with real components.

In real space the scalar product is equal to the sum of the products of corresponding components of the vectors; the length of a vector equals the square root of the sum of the squares of its components.

We shall most often have to deal with real n-dimensional space, turning to complex space only as occasion requires.

1. *Linear dependence.* A vector $Y = c_1 X_1 + c_2 X_2 + \cdots + c_m X_m$ is said to be a *linear combination* of the vectors $X_1, X_2, \ldots X_m$.

It is easily seen that if vectors $Y_1, \ldots, Y_k$ are linear combinations of the vectors $X_1, \ldots, X_m$, any linear combination $\gamma_1 Y_1 + \cdots + \gamma_k Y_k$ will also be a linear combination of the vectors $X_1, \ldots, X_m$.

Vectors $X_1, X_2, \ldots, X_m$ are called *linearly dependent* if constants $c_1, c_2, \ldots, c_m$ exist, not all zero, such that the equation

$$(5) \qquad c_1 X_1 + c_2 X_2 + \cdots + c_m X_m = 0$$

holds. If, however, this equation holds only when all the constants c_i are equal to zero, the vectors $X_1, X_2, \ldots, X_m$ are said to be *linearly independent*.

If the vectors $X_1, \ldots, X_m$ are linearly dependent, then at least one of them will be a linear combination of the rest. For if, for example, $c_m \neq 0$, we find from (5):

$$(6) \qquad X_m = -\frac{c_1}{c_m} X_1 - \cdots - \frac{c_{m-1}}{c_m} X_{m-1}.$$

THEOREM 1. *If the vectors $Y_1, \ldots, Y_k$ are linear combinations of the vectors $X_1, \ldots, X_m$, and $k > m$, the former set is linearly dependent.*

The proof will be carried through by the method of mathematical induction. For $m = 1$, the theorem is obvious. Let the theorem be true on the assumption that the number of combined vectors be $m - 1$. Under the condition of the theorem, then,

$$Y_1 = c_{11} X_1 + \cdots + c_{1m} X_m$$

$$\cdot \quad \cdot \quad \cdot \quad \cdot \quad \cdot \quad \cdot \quad \cdot \quad \cdot \quad \cdot \quad \cdot \quad \cdot$$

$$Y_k = c_{k1} X_1 + \cdots + c_{km} X_m.$$

Two cases are conceivable.

1. All the coefficients $c_{11}, \ldots, c_{k1}$ are equal to zero. Then

$Y_1, \ldots, Y_k$ are in fact linear combinations of only the $m-1$ vectors $X_2, \ldots, X_m$. On the strength of the induction hypothesis, $Y_1, \ldots, Y_k$ will be linearly dependent.

2. At least one coefficient of X_1 will be different from zero. Without violating the generality we may consider that $c_{11} \neq 0$.

Let us now consider the system of vectors

$$Y_2' = Y_2 - \frac{c_{21}}{c_{11}} Y_1$$
$$\cdots \cdots \cdots \cdots$$
$$Y_k' = Y_k - \frac{c_{k1}}{c_{11}} Y_1.$$

The vectors thus constructed are obviously linear combinations of the vectors $X_2, \ldots, X_m$, and the number of them is $k-1 > m-1$. On the strength of the induction hypothesis they are linearly dependent, i.e., constants $\gamma_2, \ldots, \gamma_k$ that are not simultaneously zero can be found such that

$$\gamma_2 Y_2' + \cdots + \gamma_k Y_k' = 0.$$

Replacing $Y_2', \ldots, Y_k'$ by their expressions in terms of $Y_1, \ldots, Y_k$, we obtain

$$\gamma_1 Y_1 + \gamma_2 Y_2 + \cdots + \gamma_k Y_k = 0,$$

where $\gamma_1 = -\dfrac{c_{21}}{c_{11}} \gamma_2 - \cdots - \dfrac{c_{k1}}{c_{11}} \gamma_k$. The numbers $\gamma_1, \ldots, \gamma_k$ are not simultaneously equal to zero and accordingly $Y_1, \ldots, Y_k$ are linearly dependent. This proves Theorem 1.

A system of linearly independent vectors is said to constitute a *basis* for a space if any vector of the space is a linear combination of the vectors of the system.

An example of a basis is the set of vectors

$$(7) \quad \begin{cases} e_1 = (1, 0, \ldots, 0) \\ e_2 = (0, 1, \ldots, 0) \\ \cdots \cdots \cdots \cdots \\ e_n = (0, 0, \ldots, 1), \end{cases}$$

for it is obvious that for any vector $X = (x_1, x_2, \ldots, x_n)$ we have

$$X = x_1 e_1 + x_2 e_2 + \cdots + x_n e_n.$$

This we shall call the *initial basis* of the space. Such a basis is not the only one possible—quite the contrary: in the choice of a basis one may be arbitrary within wide limits. Despite this, the *number* of vectors forming a basis does not depend on its selection. In proof of this, let $Y_1, \ldots, Y_k$ and $Z_1, \ldots, Z_m$ be two bases, and assume further that $k > m$. The vectors $Y_1, \ldots, Y_k$ are linear combinations of the vectors $Z_1, \ldots, Z_m$. In the light of Theorem 1, $Y_1, \ldots, Y_k$ are linearly dependent, which contradicts the definition of basis. So $k = m$. Furthermore, since the initial basis is constituted by n vectors, any other basis will also consist of n vectors. The number of vectors forming a basis thus coincides with the dimension of the space.

Let $U_1, \ldots, U_n$ form the basis of a space. Any vector X will then be a linear combination of $U_1, \ldots U_n$:

$$(8) \qquad X = \xi_1 U_1 + \xi_2 U_2 + \cdots + \xi_n U_n.$$

The coefficients of this resolution uniquely define the vector X, for if

$$X = \xi_1 U_1 + \cdots + \xi_n U_n = \xi_1' U_1 + \cdots + \xi_n' U_n,$$

then $(\xi_1 - \xi_1') U_1 + \cdots + (\xi_n - \xi_n') U_n = 0$, and accordingly

$$\xi_1 - \xi_1' = 0, \ldots, \xi_n - \xi_n' = 0,$$

in view of the linear independence of the vectors $U_1, \ldots, U_n$.

The coefficients $\xi_1, \ldots, \xi_n$ are called the *coordinates* of the vector X with respect to the basis $U_1, \ldots, U_n$. Note that the components of a vector $x_1, \ldots, x_n$ are the coordinates of the vector X with respect to the initial basis.

2. *Orthogonal systems of vectors.* The non-zero vectors of a space are said to be *orthogonal* if their scalar product equals zero. A *system* of vectors is said to be orthogonal if any two vectors of the system are orthogonal to one another. In speaking of an orthogonal system, we shall henceforth assume that all the vectors of this system are different from zero.

THEOREM 2. *The vectors forming an orthogonal system are linearly independent.*

Proof.   Let $X_1, \ldots, X_k$ be an orthogonal system, and let

$$c_1 X_1 + c_2 X_2 + \cdots + c_k X_k = 0.$$

In view of the properties of the scalar product we have:

$$
\begin{aligned}
0 &= (c_1 X_1 + \cdots + c_k X_k, X_i) \\
&= c_1(X_1, X_i) + \cdots + c_i(X_i, X_i) + \cdots + c_k(X_k, X_i) = c_i \|X_i\|^2,
\end{aligned}
$$

and, since $\|X_i\|^2 > 0$, $c_i = 0$ for any $i = 1, 2, \ldots, n$. Thus the sole possible values for $c_1, c_2, \ldots, c_n$ in the equation $c_1 X_1 + c_2 X_2 + \cdots + c_n X_n = 0$ are $c_1 = c_2 = \cdots = c_n = 0$, i.e., the vectors $X_1, X_2, \ldots, X_n$ are linearly independent.   It thence follows, first, that the number of vectors forming an orthogonal system does not exceed n, and, second, that any orthogonal system of n vectors forms a basis of the space.   Such a basis is called *orthogonal*.   If we have, in addition, $\|X_i\| = 1$, the basis is said to be *orthonormal*.   An example of an orthonormal basis is the initial basis.

From any system of linearly independent vectors $X_1, \ldots, X_k$, it is possible to go over to an orthogonal system of vectors $X_1', \ldots, X_k'$ by means of the process spoken of as *orthogonalization*.   The following theorem describes this process.

THEOREM 3. *Let $X_1, \ldots, X_n$ be linearly independent. An orthogonal system of vectors $X_1', \ldots, X_n'$ may be constructed that is connected with the original set by the relations:*

$$
(9) \quad
\begin{cases}
X_1' = X_1 \\
X_2' = X_2 + \alpha_{21} X_1 \\
\cdots \cdots \cdots \cdots \cdots \cdots \cdots \\
X_k' = X_k + \alpha_{k1} X_1 + \cdots + \alpha_{k,k-1} X_{k-1}.
\end{cases}
$$

The proof will be by induction.

Let $X_1', \ldots, X_{m-1}'$ be already constructed and different from zero. We seek X_m' in the form

$$(9') \qquad X_m' = X_m + \gamma_1 X_1' + \cdots + \gamma_{m-1} X_{m-1}'.$$

Choose the coefficients $\gamma_1, \ldots, \gamma_{m-1}$ so that $(X'_m, X'_j) = 0$ for $j = 1, \ldots, m-1$. This is easily done, for

$$(X'_m, X'_j) = (X_m, X'_j) + \gamma_j (X'_j, X'_j).$$

Now $(X'_j, X'_j) \neq 0$, since $X'_j \neq 0$ by the induction hypothesis, and it is accordingly sufficient to take

$$\gamma_j = -\frac{(X_m, X'_j)}{(X'_j, X'_j)}.$$

Replacing now $X'_1, \ldots, X'_{m-1}$ in equation (9') by their expressions in terms of $X_1, \ldots, X_{m-1}$, we obtain finally

$$X'_m = X_m + \alpha_{m1} X_1 + \cdots + \alpha_{m,m-1} X_{m-1}.$$

It remains to be proved that $X'_m \neq 0$. But this is obvious, for otherwise the vector X_m would be a linear combination of the vectors $X_1, \ldots, X_{m-1}$, which contradicts the condition of the theorem. The basis of the induction exists, since for $m = 1$ the theorem is trivial.

One may pass from any orthogonal system of vectors to the corresponding orthonormal system by dividing each vector by its length.

The process described permits of great latitude in the choice of an orthonormal basis, for one may pass from any basis to an orthonormal one by orthogonalization and normalization.

The scalar product of two vectors is very simply expressible in terms of the coordinates of these vectors with respect to any orthonormal basis, for, if $U_1, \ldots, U_n$ is an orthonormal basis, and

$$X = \xi_1 U_1 + \cdots + \xi_n U_n, \quad Y = \eta_1 U_1 + \cdots + \eta_n U_n,$$

then

$$\begin{aligned}(X, Y) &= (\xi_1 U_1 + \cdots + \xi_n U_n, \eta_1 U_1 + \cdots + \eta_n U_n) \\ &= \sum_{i=1}^{n} \sum_{j=1}^{n} (\xi_i U_i, \eta_j U_j) = \sum_{i=1}^{n} \sum_{j=1}^{n} \xi_i \bar{\eta}_j (U_i, U_j) = \sum_{i=1}^{n} \xi_i \bar{\eta}_i.\end{aligned}$$

Thus the expression of the scalar product in terms of the coordinates of the vectors with respect to any orthonormal basis coincides with its expression in terms of the components of the vectors, i.e., in terms of the coordinates with respect to the initial basis.

3. *Transformation of coordinates.* Let us elucidate the change in the coordinates of a vector that accompanies a change of basis.

Let $e_1, e_2 \ldots, e_n$ and $e'_1, e'_2, \ldots e'_n$ be two bases, and let

(10)
$$\begin{cases} e'_1 = a_{11}e_1 + a_{21}e_2 + \cdots + a_{n1}e_n \\ e'_2 = a_{12}e_1 + a_{22}e_2 + \cdots + a_{n2}e_n \\ \cdots \cdots \cdots \cdots \cdots \\ e'_n = a_{1n}e_1 + a_{2n}e_2 + \cdots + a_{nn}e_n. \end{cases}$$

We connect with the transformation of coordinates a matrix A, the columns of which consist of the coordinates of the vectors $e'_1, e'_2, \ldots e'_n$ with respect to the basis $e_1, e_2, \ldots e_n$, i.e., the matrix

(11)
$$A = \begin{bmatrix} a_{11} & a_{12} & \cdots & a_{1n} \\ a_{21} & a_{22} & \cdots & a_{2n} \\ \cdot & \cdot & \cdots & \cdot \\ a_{n1} & a_{n2} & \cdots & a_{nn} \end{bmatrix}.$$

The matrix A is non-singular, for it has an inverse, by means of which the vectors $e_1, e_2, \ldots, e_n$ are expressible in terms of the vectors $e'_1, e'_2, \ldots, e'_n$.

Now designate by $x_1, \ldots, x_n$ the coordinates of a vector X with respect to the basis $e_1, e_2, \ldots, e_n$, and by $x'_1, x'_2, \ldots, x'_n$ its coordinates with respect to the basis $e'_1, e'_2, \ldots, e'_n$. Let us determine the relation of dependence between the old and the new coordinates. We have:

$$\begin{aligned} X = x_1e_1 + x_2e_2 + \cdots + x_ne_n &= x'_1e'_1 + x'_2e'_2 + \cdots + x'_ne'_n \\ &= x'_1(a_{11}e_1 + a_{21}e_2 + \cdots + a_{n1}e_n) \\ &+ x'_2(a_{12}e_1 + a_{22}e_2 + \cdots + a_{n2}e_n) \\ &+ \cdots \cdots \cdots \cdots \cdots \\ &+ x'_n(a_{1n}e_1 + a_{2n}e_2 + \cdots + a_{nn}e_n) \\ &= (a_{11}x'_1 + a_{12}x'_2 + \cdots + a_{1n}x'_n)e_1 \\ &+ (a_{21}x'_1 + a_{22}x'_2 + \cdots + a_{2n}x'_n)e_2 \\ &+ \cdots \cdots \cdots \cdots \cdots \\ &+ (a_{n1}x'_1 + a_{n2}x'_2 + \cdots + a_{nn}x'_n)e_n, \end{aligned}$$

whence, on the strength of the linear independence of the vectors $e_1, e_2, \ldots, e_n$:

$$(12) \quad \begin{cases} x_1 = a_{11}x'_1 + a_{12}x'_2 + \cdots + a_{1n}x'_n \\ x_2 = a_{21}x'_1 + a_{22}x'_2 + \cdots + a_{2n}x'_n \\ \cdot \; \cdot \; \cdot \; \cdot \; \cdot \; \cdot \; \cdot \; \cdot \; \cdot \; \cdot \; \cdot \\ x_n = a_{n1}x'_1 + a_{n2}x'_2 + \cdots + a_{nn}x'_n. \end{cases}$$

The last equations may be written in matrix form. Of course the set of a vector's coordinates can be considered either as a column or as a row. In view of the definition of matrix multiplication, we can postmultiply a square matrix only by a column, not by a row. In future we shall, therefore (except by special stipulation), take the coordinates of a vector as a *column*. Often in arguments where the basis is fixed (for instance, when the vector is given in terms of its coordinates with respect to the initial basis), we shall identify the vector with the column of its coordinates.

Equation (12) may be written in the form

$$(13) \qquad\qquad x = Ax',$$

where

$$(14) \qquad x = \begin{bmatrix} x_1 \\ x_2 \\ \vdots \\ x_n \end{bmatrix} \quad \text{and} \quad x' = \begin{bmatrix} x'_1 \\ x'_2 \\ \vdots \\ x'_n \end{bmatrix}$$

are the coordinate columns of the vector X with respect to the bases $e_1, \ldots, e_n$ and $e'_1, \ldots, e'_n$ respectively.

4. *Subspaces.* A set of vectors $X \subset R_n$ such that any linear combination of the vectors of this set is itself a vector of the same set, is said to be a *subspace* of the space R_n. If a group of vectors $U_1, \ldots, U_m$ that are linearly independent—or even linearly dependent—be given, then the set of all possible linear combinations of them will obviously constitute a subspace. A subspace constructible in this manner is said to be the subspace *spanned* by the system of vectors $U_1, \ldots, U_m$.

We shall show that a basis exists in every subspace, i.e., that there is a set of linearly independent vectors, by linear combinations of which one may exhaust the entire subspace. Let us construct the basis in the following manner. Take first an arbitrary vector X_1, different from zero, and consider all its linear combinations, i.e., all vectors of the form cX_1. If these exhaust the entire subspace, X_1 then forms its basis. If the contrary is true, a vector X_2 will be found, linearly independent of X_1. Consider the set of linear combinations of X_1 and X_2: if they do not exhaust the subspace, a vector will be found linearly independent of them, and so forth. The process cannot go on interminably, for in the space R_n there cannot be more than n linearly independent vectors. We shall thus have constructed a finite system of vectors $X_1, \ldots, X_k$ such that their linear combinations exhaust our subspace, i.e., we shall have constructed our basis.

It will be remarked that the reasoning set forth indicates much latitude in choice of a basis. However the *number* of vectors forming a basis will not depend on the manner of its selection, in the light of Theorem 1. That number is called the *dimension* of the subspace.

Note that the set composed solely of the null vector, as also the set composed of the entire space, will each be subspaces in the sense of our definition. We shall regard them as trivial subspaces.

5. *The connection between the dimension of a subspace and the rank of a matrix.* We introduce the important concept of rank, appropriate to any rectangular matrix A,

$$A = \begin{bmatrix} a_{11} & a_{12} & \cdots & a_{1n} \\ a_{21} & a_{22} & \cdots & a_{2n} \\ \cdot & \cdot & \cdots & \cdot \\ a_{m1} & a_{m2} & \cdots & a_{mn} \end{bmatrix}.$$

Any determinant whose rows and columns "fit" the rows and columns of a matrix is called a *minor* of this matrix. More exactly, a *minor* of order k of the matrix A is a determinant of the kth order formed from the elements situated at the intersections of any k rows and any k columns of the matrix A, in their natural arrangement.

The order of whatever non-vanishing minor is of largest order is called the *rank* of the matrix A. In other words, the rank of a matrix is a number r such that among the minors of the matrix there exists a non-zero minor of order r, but all minors of order $r+1$ and higher are equal to zero or are not composible (as in the case, for instance, of a rectangular $m \times r$ matrix, $m > r$).

The following important theorem is valid:

THEOREM. *The maximum number of linearly independent rows of a matrix, as also the maximum number of linearly independent columns, coincides with the rank of the matrix.*

From this theorem it follows directly that the dimension of the subspace spanned by the vectors $U_1, \ldots, U_m$ equals the rank of the matrix composed of the components of these vectors.

Indeed, if the rank of a matrix whose columns are the components of the vectors $U_1, \ldots, U_m$ equals r, then of these m vectors r will be linearly independent, and these will correspond to the linearly independent columns of the matrix; all the rest of the columns will be linear combinations of them. Any vector subspace is a linear combination of the vectors $U_1, \ldots, U_m$, which are themselves linear combinations of but r selected linearly independent vectors. Consequently any vector is a linear combination of r vectors, and therefore the rank r of the matrix in question coincides with the dimension of the subspace.

§ 3. LINEAR TRANSFORMATIONS

1. Let us associate with each vector X of a space a certain vector Y of the same space. Such an association we shall call a *transformation* of the space. We shall designate the result of the application of transformation A to the vector X by AX.

We shall call the transformation A *linear* if

1. $A(\alpha X) = \alpha AX$, for any complex number α;

2. $A(X_1 + X_2) = AX_1 + AX_2$.

We shall define, furthermore, operations upon linear transformations. The *product* of the linear transformations A and B, $AB = C$, will be a transformation constituted by the transformations B and A in turn, B being completed first, and then A.

The product of linear transformations is a linear transformation, as is readily seen, since

$$C(X_1 + X_2) = A(B(X_1 + X_2)) = A(BX_1 + BX_2)$$

(1)
$$= ABX_1 + ABX_2 = CX_1 + CX_2,$$

$$C\alpha X = AB\alpha X = A\alpha BX = \alpha ABX = \alpha CX.$$

The *sum* of the linear transformations A and B will be a transformation C which associates the vector X with the vector $AX + BX$. This sum of linear transformations is obviously itself a linear transformation.

2. *Representation of a linear transformation by a matrix.* Let us choose, in the space R_n, some basis $e_1, e_2, \ldots, e_n$. A linear transformation relates to the vectors of the basis the vectors Ae_1, $Ae_2, \ldots, Ae_n$.

Let $Ae_1, \ldots, Ae_n$ be given in terms of their coordinates with respect to the basis $e_1, e_2, \ldots e_n$, i.e., let

$$(2) \quad \begin{cases} Ae_1 = a_{11}e_1 + a_{21}e_2 + \cdots + a_{n1}e_n \\ Ae_2 = a_{12}e_1 + a_{22}e_2 + \cdots + a_{n2}e_n \\ \cdot \quad \cdot \quad \cdot \quad \cdot \quad \cdot \quad \cdot \quad \cdot \quad \cdot \quad \cdot \quad \cdot \quad \cdot \quad \cdot \\ Ae_n = a_{1n}e_1 + a_{2n}e_2 + \cdots + a_{nn}e_n. \end{cases}$$

Consider the matrix A, its columns composed of the coordinates of the vectors $Ae_1, Ae_2, \ldots, Ae_n$:

$$(3) \quad A = \begin{bmatrix} a_{11} & a_{12} & \cdots & a_{1n} \\ a_{21} & a_{22} & \cdots & a_{2n} \\ \cdot & \cdot & \cdots & \cdot \\ a_{n1} & a_{n2} & \cdots & a_{nn} \end{bmatrix}.$$

We shall show that the matrix A uniquely defines the linear transformation.[1]

Indeed, if the matrix A is known for the linear transformation, i.e., if Ae_1, Ae_2, ... Ae_n are determined, this is sufficient to find the transformation of any vector, for if

$$X = x_1 e_1 + \cdots + x_n e_n,$$

then

$$AX = x_1 Ae_1 + \cdots + x_n Ae_n.$$

Hence the coordinates of the transformed vector are easily found, for we have

$$Y = AX = \sum_{k=1}^{n} y_k e_k \doteq \sum_{i=1}^{n} x_i A e_i = \sum_{i=1}^{n} \sum_{k=1}^{n} a_{ki} x_i e_k,$$

whence

$$y_k = \sum_{i=1}^{n} a_{ki} x_i,$$

or, in matrix notation,

$$(4) \qquad y = Ax,$$

where y and x are columns of the coordinates of vectors Y and X.

Conversely, an arbitrary matrix A may be connected with a certain linear transformation. Indeed, the transformation given by the formula

$$y = Ax,$$

where y and x are, as above, the columns of coordinates of the vectors Y and X, is linear for any matrix A.

The established one-to-one correspondence between transformations and matrices is preserved when operations are performed upon transformations, for *the matrix of the sum of transformations equals the sum of the matrices of the summand transformations*, and *the matrix of a product of transformations equals the product of the matrices corresponding to the factor transformations*.

3. *The connection between the matrices of a linear transformation with respect to different bases.* We will now elucidate how the matrix of a linear transformation changes with a change of the basis of the space.

---

[1] Note, however, that the matrix of the coefficients in the relations (2) forms a matrix which is the transpose of that that we connect with the linear transformation.

Assume that from the basis $e_1, \ldots, e_n$ we have passed to the basis $e'_1, \ldots, e'_n$, and let

$$e'_1 = c_{11}e_1 + c_{21}e_2 + \cdots + c_{n1}e_n$$
$$e'_2 = c_{12}e_1 + c_{22}e_2 + \cdots + c_{n2}e_n$$
$$\cdot \quad \cdot \quad \cdot \quad \cdot \quad \cdot \quad \cdot \quad \cdot \quad \cdot \quad \cdot \quad \cdot \quad \cdot \quad \cdot$$
$$e'_n = c_{1n}e_1 + c_{2n}e_2 + \cdots + c_{nn}e_n.$$

The coordinates of any vector of the space will have been changed accordingly by the formula

$$x = Cx',$$

where

$$C = \begin{bmatrix} c_{11} & c_{12} & \cdots & c_{1n} \\ c_{21} & c_{22} & \cdots & c_{2n} \\ \cdot & \cdot & \cdot & \cdot \\ c_{n1} & c_{n2} & \cdots & c_{nn} \end{bmatrix}, \quad x = \begin{bmatrix} x_1 \\ x_2 \\ \vdots \\ x_n \end{bmatrix}, \quad x' = \begin{bmatrix} x'_1 \\ x'_2 \\ \vdots \\ x'_n \end{bmatrix}.$$

The matrix of the transition, C, is evidently non-singular. It will coincide with the matrix of the linear transformation sending the basis $e_1, e_2, \ldots, e_n$ into the basis $e'_1, e'_2, \ldots e'_n$.

Let us now consider a linear transformation A, and let the matrix A correspond to it with respect to the basis $e_1, e_2, \ldots, e_n$, and the matrix B with respect to the basis $e'_1, e'_2, \ldots, e'_n$.

If x is the column of the coordinates of the vector X with respect to the basis $e_1, \ldots, e_n$, and x' that with respect to the basis $e'_1, \ldots, e'_n$, y and y' being the analogous columns for vector Y, we have

$$y = Ax$$
$$y' = Bx'.$$

But $x = Cx'$, $y = Cy'$, and therefore

$$Cy' = ACx'$$

or

$$y' = Bx' = C^{-1}ACx'.$$

Thus similar matrices correspond to the same linear transformation with respect to different bases. Furthermore, the matrix by means of which the similarity transformation is effected coincides with the matrix of transformation of coordinates.

4. *The transfer rule for a matrix in a scalar product.* Let X and Y be two vectors given by their components with respect to the initial basis: $X = (x_1, \ldots, x_n)$, $Y = (y_1, \ldots, y_n)$, and let A be a linear transformation with matrix $A = (a_{ik})$. Designate by A^* the linear transformation with matrix A^*, the elements of which are the complex-conjugates of their counterparts in A, and which are placed in transposed positions: $A^* = (a_{ik})^* = (\overline{a_{ik}'}) = (\overline{a_{ki}})$. We shall prove the following formula:

$$(AX, Y) = (X, A^*Y).$$

In demonstration, we have

$$(AX, Y) = \sum_{i=1}^{n} \sum_{k=1}^{n} a_{ik} x_k \bar{y}_i = \sum_{k=1}^{n} x_k \overline{\sum_{i=1}^{n} \bar{a}_{ik} y_i} = (X, A^*Y).$$

5. *The rank of a linear transformation.* Let A be a certain linear transformation. The set of vectors AX will obviously constitute a subspace, which we shall denote by AR_n.

The dimension of this subspace is said to be the *rank* of the *transformation* A.

We shall show the rank of a transformation to be equal to the rank of the matrix corresponding to this transformation on any basis whatever, $e_1, e_2, \ldots, e_n$. Obviously the subspace AR_n is spanned by the vectors $Ae_1, Ae_2, \ldots, Ae_n$. The dimension of AR_n is accordingly equal to the rank of a matrix whose columns are composed of the coordinates of the vectors $Ae_1, Ae_2, \ldots, Ae_n$, i.e., to the rank of a matrix corresponding to the transformation.

Since the dimension of a subspace does not depend upon the selection of the basis, it follows from the foregoing that the ranks of similar matrices are equal.

6. *The proper vectors of a linear transformation.* By a *proper vector* (*characteristic vector*, *latent vector* or *eigenvector*) of a linear transformation A is meant any non-zero vector X such that

(6) $$AX = \lambda X,$$

where λ is any complex number.

The number λ is called a *proper number* (*characteristic number, latent root, or eigenvalue*) of the transformation. The *spectrum* of the transformation is the aggregate of its proper numbers.

The proper numbers and proper vectors of the transformation may be determined in the following manner. Let the transformation A be connected with the matrix $A = (a_{ik})$ with respect to some basis; let the coordinates of the proper vector X, with respect to this basis, be $x_1, \ldots, x_n$.

The coordinates of the vector AX will then be:

$$\left[\sum_{k=1}^{n} a_{1k}x_k, \ldots, \sum_{k=1}^{n} a_{nk}x_k \right],$$

and thus for the determination of $x_1, x_2, \ldots, x_n$ and the proper number λ we will have the system of equations:

$$(7) \quad \begin{cases} a_{11}x_1 + a_{12}x_2 + \cdots + a_{1n}x_n = \lambda x_1 \\ a_{21}x_1 + a_{22}x_2 + \cdots + a_{2n}x_n = \lambda x_2 \\ \cdots \cdots \cdots \cdots \cdots \cdots \cdots \\ a_{n1}x_1 + a_{n2}x_2 + \cdots + a_{nn}x_n = \lambda x_n \end{cases}$$

or

$$(7') \quad \begin{cases} (a_{11} - \lambda)x_1 + a_{12}x_2 + \cdots + a_{1n}x_n = 0 \\ a_{21}x_1 + (a_{22} - \lambda)x_2 + \cdots + a_{2n}x_n = 0 \\ \cdots \cdots \cdots \cdots \cdots \cdots \cdots \\ a_{n1}x_1 + a_{n2}x_2 + \cdots + (a_{nn} - \lambda)x_n = 0. \end{cases}$$

This system of homogeneous equations in $x_1, \ldots, x_n$ will have a non-zero solution only in case

$$\begin{vmatrix} a_{11} - \lambda & a_{12} & \cdots & a_{1n} \\ a_{21} & a_{22} - \lambda & \cdots & a_{2n} \\ \cdots & \cdots & \cdots & \cdots \\ a_{n1} & a_{n2} & \cdots & a_{nn} - \lambda \end{vmatrix} = 0,$$

i.e., if λ is a zero of the characteristic polynomial of the matrix. Thus the following is valid:

THEOREM. *The proper numbers of a transformation coincide with the zeros of the characteristic polynomial of a matrix that is connected with this transformation with respect to an arbitrary basis.*

From the theorem known as the fundamental theorem of higher algebra, we know that every polynomial has at least one zero. A linear transformation will consequently have at least one proper number, which may be complex even though the matrix of the transformation be real. In view of the theory of linear homogeneous systems of equations, there will be a non-zero solution of system (7) for each proper number, i.e., with each proper number at least one proper vector is associated.

Obviously if X is a proper vector of the transformation A, then, for all $c \neq 0$, cX will also be a proper vector of transformation A corresponding to the same proper number. Furthermore if several proper vectors correspond to some one proper number, then any linear combination of them will be a proper vector of the transformation associated with the same number. The set of proper vectors corresponding to the same proper number forms a linear subspace. We shall establish that its dimension, l, does not exceed the multiplicity of the proper number. Indeed, let $X_1, \ldots, X_l$ be linearly independent proper vectors corresponding to the same proper number λ_1. Construct a basis of the space $X_1, \ldots, X_n$, having taken as the first l vectors the vectors $X_1, \ldots, X_l$. With respect to this basis the linear transformation under consideration is connected with a matrix whose first l columns have the form

(7'b)
$$
\begin{matrix}
\lambda_1 & 0 & \ldots & 0 \\
0 & \lambda_1 & \ldots & 0 \\
\cdot & \cdot & \cdot & \cdot \cdot \cdot \\
0 & 0 & \ldots & \lambda_1 \\
\cdot & \cdot & \cdot & \cdot \cdot \cdot \\
0 & 0 & \ldots & 0 \,,
\end{matrix}
$$

for $AX_1 = \lambda_1 X_1, \ldots, AX_l = \lambda_1 X_l$. Now $(\lambda - \lambda_1)^l$ is a factor of the characteristic polynomial of this matrix, and accordingly λ_1 is of multiplicity k not less than l, i.e., $l \leqslant k$. It would naturally be supposed that $l = k$, i.e., that to a k-multiple root of the characteristic polynomial there correspond k linearly independent proper vectors. But this is in fact not true.   In reality, the number of linearly independent vectors may be less than the multiplicity of the proper number.

Let us confirm the preceding statement with an example.   Consider the linear transformation with the matrix

$$A = \begin{bmatrix} 3 & 1 \\ 0 & 3 \end{bmatrix}.$$

Then $|A - \lambda I| = (\lambda - 3)^2$, and thus $\lambda = 3$ is a double root of the characteristic polynomial.

The system of equations for determining the coordinates of the proper vector of the transformation A will be:

$$3x_1 + x_2 = 3x_1$$
$$3x_2 = 3x_2$$

whence $x_2 = 0$, and thus all the proper vectors of the transformation in question will be $(x_1, 0) = x_1(1, 0)$.   So in this instance only one linearly independent vector is associated with a double root.

Generally speaking, the coordinates of a proper vector on the chosen basis are to be determined from the system (7) of linear equations, in which for λ the proper number λ_i is substituted.   But as is known from the theory of systems of linear equations, the number of linearly independent solutions of a homogeneous system equals $n - r$, where r is the rank of the matrix composed of the coefficients of the system.   Therefore if r denotes the rank of the matrix $A - \lambda I$, $l = n - r$.   Thus $n - r \leqslant k$, and the equality does not always hold.[1]

In case the basis does not change in the course of the argument, we shall often identify the linear transformation with the matrix of the linear transformation with respect to this basis, and any vector of the space with the column of its coordinates.

On this agreement, it makes sense to speak of a *proper vector* of a *matrix*, understanding by this a column x satisfying the condition

$$Ax = \lambda x.$$

We remark that if a proper number of a real matrix is complex,

[1] V. I. Smirnov, [1]; A. G. Kurosh, [1].

the coordinates of an associated proper vector will be complex. A vector whose coordinates are the complex conjugates of those of a given proper vector of a real matrix is also a proper vector of that matrix, and is associated with the complex conjugate proper number. To convince oneself of this, it is enough to change all numbers in the equation $Ax = \lambda x$ into their complex conjugates.

7. *Properties of the proper numbers and proper vectors of a matrix.* We shall establish several properties of the proper numbers and proper vectors of a real matrix.

First of all we note that a matrix and its transpose have identical characteristic polynomials and consequently identical spectra. This is evident since $|A' - \lambda I| = |A - \lambda I|$, on the strength of the fact that a determinant is not altered when its rows and columns are interchanged.

Now let λ_r and λ_s denote distinct proper numbers of the real matrix A, and $\bar{\lambda}_s$ the complex conjugate of λ_s. As we saw above, $\bar{\lambda}_s$ is also a proper number of matrix A, and thus also of the transposed matrix A'. Let X_r be the proper vector of the matrix A belonging to the proper number λ_r, and X'_s the proper vector of the matrix A' belonging to the proper number $\bar{\lambda}_s$. We shall show that X_r and X'_s are orthogonal.

With this object in view, let us form the scalar product (AX_r, X'_s) and reckon it by two methods.

By one method we have

$$(AX_r, X'_s) = (\lambda_r X_r, X'_s) = \lambda_r(X_r, X'_s).$$

On the other hand, since matrix A is real, we have $A^* = A'$, and therefore

$$(AX_r, X'_s) = (X_r, A'X'_s) = (X_r, \bar{\lambda}_s X'_s) = \lambda_s(X_r, X'_s).$$

Thus $\lambda_r(X_r, X'_s) = \lambda_s(X_r, X'_s)$. But the condition was that $\lambda_r \neq \lambda_s$, and therefore $(X_r, X'_s) = 0$, which is what was required to be proved.

In case all the proper numbers are distinct, the demonstrated property gives $n^2 - n$ relations of orthogonality between the proper vectors of matrices A and A'. We shall later return to these properties in more detail.

For a *real symmetric* matrix, the properties of orthogonality are considerably simplified, thanks to the fact that *all its proper numbers are real*. In proof, letting λ and X be respectively proper number and vector, we have $(AX, X) = \lambda(X, X)$; $(AX, X) = (X, A'X) = (X, AX) = (X, \lambda X) = \bar{\lambda}(X, X)$. Thus $(\lambda - \bar{\lambda})(X, X) = 0$, and as $(X, X) > 0$, $\lambda = \bar{\lambda}$, i.e., λ is real.

From the reality of the proper numbers of a real symmetric matrix it follows that vectors with real components may be taken as the proper vectors belonging to those roots; the components will indeed be found by solving the linear homogeneous system with real coefficients.

The orthogonality property of the proper vectors of a real symmetric matrix is very simply formulated in view of the coincidence of the matrix with its transpose and the reality of the proper numbers, viz.: *proper vectors belonging to distinct proper numbers are orthogonal*.

8. *The proper numbers of a positive-definite quadratic form.* A homogeneous polynomial of the second degree in several variables $x_1, \ldots, x_n$ is called a *quadratic form*. We shall consider only those with real coefficients. Any quadratic form may be written as

$$\Phi(x_1, x_2, \ldots, x_n) = \sum_{i, k=1}^{n} a_{ik} x_i x_k,$$

where $a_{ik} = a_{ki}$.

A quadratic form is said to be *positive-definite* if its values are positive for any real values of $x_1, \ldots, x_n$, not all zero simultaneously.

It is evident that the diagonal coefficients of a positive-definite form are positive, for

$$a_{11} = \Phi(1, 0, \ldots, 0), \quad a_{22} = \Phi(0, 1, \ldots, 0), \ldots$$

$$a_{nn} = \Phi(0, 0, \ldots, 1).$$

Denoting by X the vector with components $(x_1, \ldots, x_n)$, we may write a quadratic form as (AX, X), A being the matrix composed of the coefficients of the form. This matrix is symmetric, on the strength of the definition. The proper numbers of the matrix are called the proper numbers of the quadratic form. In view of the previous results, *all proper numbers of a quadratic form are real*.

We shall show that *if a quadratic form is positive-definite, its proper numbers are positive*.

In demonstration, let X be a real proper vector, belonging to λ, a proper number of the matrix of the form. Then, since the form is positive-definite, $(AX, X) > 0$. On the other hand, $(AX, X) = \lambda(X, X)$. Thus

$$\lambda = \frac{(AX, X)}{(X, X)}.$$

But both numerator and denominator of this fraction are positive, and consequently $\lambda > 0$, which is what was required to be proved.

Let there now be given any real, non-singular matrix A. Obviously $B = A'A$ is a symmetric matrix, since $B' = (A'A)' = A'A'' = A'A = B$.

We shall show that a quadratic form with matrix B is positive-definite. We have, indeed,

$$(BX, X) = (A'AX, X) = (AX, AX) > 0$$

for any real vector X.

We shall establish, lastly, that if A is the matrix of a positive-definite quadratic form, $(AX, X) > 0$ even for a complex vector X.

In proof, let $X = Y + iZ$, where Y and Z are vectors with real components. Then

$$(AX, X) = (AY + iAZ, Y + iZ)$$
$$= (AY, Y) + i(AZ, Y) - i(AY, Z) + (AZ, Z)$$
$$= (AY, Y) + (AZ, Z) > 0$$

because $(AZ, Y) = (Z, AY) = (AY, Z)$.

In complex space, instead of the quadratic form one deals with an *Hermitian* form, an expression of the type

$$\sum_{i,\,k=1}^{n} a_{ik} x_i \bar{x}_k,$$

under the condition that $a_{ki} = \bar{a}_{ik}$.

The matrix of an Hermitian form is called *Hermitian* (or *Hermitian symmetric*); a linear transformation with an Hermitian matrix relative to an orthonormal base is called *self-conjugate*. It is obvious that

$$\sum a_{ik} x_i \bar{x}_k = (AX, X).$$

To show that all the values of an Hermitian form are real, we have only to note that

$$(AX, X) = (X, A^*X) = \overline{(AX, X)}.$$

If all the values of an Hermitian form are positive, it is called *positive-definite*.

It can be shown that the *proper numbers of an Hermitian matrix are real. The proper numbers of a positive-definite Hermitian form are positive.*

9. *The reduction of a matrix to diagonal form.* Let us consider the matrix A all of whose proper numbers, $\lambda_1, \ldots, \lambda_n$, are distinct, and the transformation A connected with it with respect to the initial basis. It will have n distinct proper vectors $X_1, \ldots, X_n$. We shall show that *the vectors $X_1, \ldots, X_n$ are linearly independent.*

Assume the contrary: let the vectors $X_1, \ldots, X_n$ be linearly dependent. Without detriment to the generality we may assume that the vectors $X_1, \ldots, X_k$, where $k < n$, are linearly independent, and thus that the vectors $X_{k+1}, \ldots, X_n$ are linear combinations of them. In particular, let

$$(8) \qquad X_n = \sum_{i=1}^{k} c_i X_i;$$

then

$$AX_n = A \sum_{i=1}^{k} c_i X_i = \sum_{i=1}^{k} \lambda_i c_i X_i.$$

On the other hand,

$$AX_n = \lambda_n X_n = \sum_{i=1}^{k} \lambda_n c_i X_i,$$

whence

$$(9) \qquad \sum_{i=1}^{k} (\lambda_n - \lambda_i) c_i X_i = 0.$$

But $\lambda_n \neq \lambda_i$, on assumption. Thus, since the vectors X_i are linearly independent, all the coefficients c_i equal zero, the therefore $X_n = 0$, which contradicts the definition of a proper vector. So the vectors $X_1, X_2, \ldots, X_n$ are linearly independent. Let us adopt them as a new basis of the space. With respect to the new basis the linear transformation A will be connected with a matrix whose columns are composed of the coordinates of the vectors AX_1, AX_2, ..., AX_n with respect to the basis X_1, X_2, ..., X_n.

But

$$AX_k = \lambda_k X_k,$$

and the matrix of the transformation on the new basis will consequently be diagonal: $\ulcorner \lambda_1, \lambda_2, \ldots \lambda_{n\ \lrcorner}$.

So the linear transformation A has, with respect to the initial basis, the matrix A, and with respect to the basis of the proper vectors, the diagonal matrix $\ulcorner \lambda_1, \lambda_2, \ldots \lambda_{n\ \lrcorner}$. Accordingly, on the strength of what has been noted above,

(10) $$V^{-1}AV = \ulcorner \lambda_1, \lambda_2, \ldots, \lambda_{n\ \lrcorner},$$

where V is the matrix whose columns are the coordinates (with respect to the initial basis) of the proper vectors.

Observation. If the proper numbers of a matrix are of multiplicity greater than one, but to each proper number there correspond as many proper vectors as it has multiplicity, the matrix may also be reduced to diagonal form. This will be the case, for example, with symmetric matrices: it can be proved that *to each proper number of a symmetric matrix there correspond as many linearly independent proper vectors as the multiplicity of the proper number.* Moreover, the linearly independent proper vectors belonging to a single proper number may be subjected to the orthogonalizing process. We have seen, too, that the proper vectors of a symmetric matrix that belong to distinct proper numbers are mutually orthogonal. Thus for a symmetric matrix it is possible to construct an orthogonal system of proper vectors forming a basis for the whole space.

The question of the transformation of a symmetric matrix to diagonal form is closely connected with the theory of quadratic forms.

10. *The proper numbers and proper vectors of similar matrices.* It has been established that similar matrices have identical characteristic polynomials, and consequently identical spectra of proper numbers.

We have explained the geometrical cause of this circumstance, viz.: similar matrices may be regarded as matrices of one and the same transformation, referred to different bases. Therefore the proper vectors of similar matrices are the columns of the coordinates of the proper vectors of the transformation under consideration, with respect to different bases, and are thus connected by the relation $x' = C^{-1}x$, C being the matrix of transformation of coordinates. This circumstance may be verified formally: if $Ax = \lambda x$, $(C^{-1}AC)(C^{-1}x) = \lambda(C^{-1}x)$.

11. *The proper numbers of a polynomial in a matrix.* Let A be a matrix with proper numbers $\lambda_1, \ldots, \lambda_n$, and let $\varphi(x) = a_0 + a_1 x + \cdots + a_m x^m$ be the given polynomial. Then the proper numbers of the matrix $\varphi(A)$ will be $\varphi(\lambda_1), \varphi(\lambda_2), \ldots, \varphi(\lambda_n)$.

This is readily established for a matrix all of whose proper numbers are distinct. Indeed, such a matrix can be reduced to diagonal form by a similarity transformation:

$$A = C^{-1} \ulcorner \lambda_1, \lambda_2, \ldots, \lambda_n \urcorner C.$$

Accordingly,

$$\varphi(A) = C^{-1} \varphi \ulcorner \lambda_1, \lambda_2, \ldots, \lambda_n \urcorner C.$$

But

$$\varphi(\ulcorner \lambda_1, \ldots, \lambda_n \urcorner) = \ulcorner \varphi(\lambda_1), \varphi(\lambda_2), \ldots, \varphi(\lambda_n) \urcorner,$$

which follows from the fact that

$$\ulcorner \lambda_1, \ldots, \lambda_n \urcorner^k = \ulcorner \lambda_1^k, \ldots, \lambda_n^k \urcorner.$$

Consequently

$$\varphi(\ulcorner \lambda_1, \ldots, \lambda_n \urcorner) = \sum_{k=0}^{m} a_k \ulcorner \lambda_1^k, \ldots, \lambda_n^k \urcorner$$

$$= \ulcorner \sum_{k=0}^{m} a_k \lambda_1^k, \ldots, \sum_{k=0}^{m} a_k \lambda_n^k \urcorner$$

$$= \ulcorner \varphi(\lambda_1), \varphi(\lambda_2), \ldots, \varphi(\lambda_n) \urcorner.$$

Thus the matrix $\varphi(A)$ is similar to the matrix $\ulcorner \varphi(\lambda_1), \ldots, \varphi(\lambda_n) \urcorner$ and accordingly its proper numbers are $\varphi(\lambda_1), \varphi(\lambda_2), \ldots, \varphi(\lambda_n)$, Q.E.D.

This result remains true for any matrix, of which one may readily convince oneself, for example, by considerations of continuity.

We particularly note that the proper numbers of the matrix A^k are λ^k.

12. *The normalization of the proper vectors of a matrix. The second group of orthogonality relations.* Let A be a real matrix whose proper

numbers, $\lambda_1, \lambda_2, \ldots, \lambda_n$, are distinct, and let $X_1, X_2, \ldots, X_n$ be the proper vectors corresponding to them. As we saw, the transposed matrix, A', has the same proper numbers. Let $X'_1, X'_2, \ldots, X'_n$, be the proper vectors of the matrix A', and their enumeration be so chosen that X_i and X'_i belong to complex-conjugate proper numbers. We established above that the following orthogonality relation holds: $(X'_i, X_j) = 0$ for $i \neq j$. We shall show now that, having chosen $X'_1, X'_2, \ldots, X'_n$ in any manner (they are determined but for a numerical multiplier), we may normalize the vectors $X_1, X_2, \ldots, X_n$ so that $(X'_i, X_i) = 1$.

In demonstration of this, the vectors $X_1, X_2, \ldots, X_n$ are known to be linearly independent, and they accordingly form a basis of the space. Resolve X'_i in terms of this basis:

$$X'_i = \gamma_1 X_1 + \gamma_2 X_2 + \cdots + \gamma_n X_n.$$

Forming the scalar product (X'_i, X'_i), we obtain

$$\begin{aligned}(X'_i, X'_i) &= \gamma_1 (X'_i, X_1) + \cdots + \gamma_i (X'_i, X_i) + \cdots + \gamma_n (X'_i, X_n) \\ &= \gamma_i (X'_i, X_i),\end{aligned}$$

whence we conclude that $(X'_i, X_i) = \alpha_i \neq 0$, for $(X'_i, X'_i) > 0$.

Adopting instead of the vectors $X_1, \ldots, X_n$ the vectors $\dfrac{1}{\bar{\alpha}_1} X_1, \ldots, \dfrac{1}{\bar{\alpha}_n} X_n$, we arrive at the required normalization, since

$$\left(X'_i, \frac{1}{\bar{\alpha}_i} X_i\right) = \frac{1}{\alpha_i} (X'_i, X_i) = 1.$$

From the relations of orthogonality and normality set forth above, we may derive another group of relations between the components of the proper vectors of the matrix A and its transpose.

Form the matrices

$$X' = \begin{bmatrix} x'_{11} & x'_{21} & \cdots & x'_{n1} \\ x'_{12} & x'_{22} & \cdots & x'_{n2} \\ \cdots & \cdots & \cdots & \cdots \\ x'_{1n} & x'_{2n} & \cdots & x'_{nn} \end{bmatrix} \text{ and } X = \begin{bmatrix} x_{11} & x_{12} & \cdots & x_{1n} \\ x_{21} & x_{22} & \cdots & x_{2n} \\ \cdots & \cdots & \cdots & \cdots \\ x_{n1} & x_{n2} & \cdots & x_{nn} \end{bmatrix}.$$

The columns of the matrix X are composed of the components of the vectors $X_1, \ldots, X_n$. The rows of the matrix X' are composed of the numbers complex-conjugate with the components of the vectors $X'_1, \ldots, X'_n$. (We observe that the numbers that are the complex conjugates of the components of the vectors $X'_1, \ldots, X'_n$ are the components of the vectors $\overline{X'_1}, \ldots, \overline{X'_n}$, which will also be proper vectors of the matrix A' belonging to the proper numbers $\lambda_1, \ldots, \lambda_n$. Thus the ith row of the matrix X' and the ith column of the matrix X are composed of the components of proper vectors of matrices A' and A, belonging to the same proper number λ_i, and not to proper numbers that are complex conjugates of each other.[1])

It is readily seen that:

$$(11) \qquad X'X = I,$$

for we have the element of the ith row and jth column of the matrix $X'X$ equalling $\sum_{k=1}^{n} x'_{ki} x_{kj} = (X_j, X'_i) = \delta_{ij}$, where δ_{ij} is Kronecker's delta:

$$\delta_{ij} = \begin{array}{ll} 0, & i \neq j; \\ 1, & i = j. \end{array}$$

Thus X' and X are mutually inverse matrices, and accordingly XX' is likewise equal to I. This gives a second group of orthogonality and normality relations between the proper vectors of the matrices A and A', viz.:

$$(12) \qquad \begin{cases} x_{i1}x'_{j1} + x_{i2}x'_{j2} + \cdots + x_{in}x'_{jn} = 0 & (i \neq j) \\ x_{i1}x'_{i1} + x_{i2}x'_{i2} + \cdots + x_{in}x'_{in} = 1. \end{cases}$$

[1] [This may perhaps be clarified by a summary in matrix notation. By the definitions at the beginning of Paragraph 12,

$$AX_i = \lambda_i X_i, \qquad (1)$$
$$A'X'_i = \lambda_i X'_i. \qquad (2)$$

From (2),

$$\overline{A'X'_i} = \overline{\lambda_i X'_i}, \qquad (2a)$$

i.e., $\qquad A'\overline{X'_i} = \lambda_i \overline{X'_i}, \quad$ for A is real. (2b)

Now by definition of the matrix X',

$$X' = \begin{bmatrix} X'_1 \\ X'_2 \\ \vdots \\ X'_n \end{bmatrix}.$$

From which, with (2b) and (1) in view, the author's statement is confirmed.]

Thus for a matrix of the second order the ordinary conditions of orthogonality may be written in the form (we preserve only one index of the components of the proper vectors, designating the first components by x, the second by y):

$$x_1 x_2' + y_1 y_2' = 0$$
$$x_2 x_1' + y_2 y_1' = 0,$$

and the normality conditions:

$$x_1' x_1 + y_1' y_1 = 1$$
$$x_2' x_2 + y_2' y_2 = 1.$$

The new relations will be

$$x_1 y_1' + x_2 y_2' = 0 \qquad x_1 x_1' + x_2 x_2' = 1$$
$$y_1 x_1' + y_2 x_2' = 0 \qquad y_1 y_1' + y_2 y_2' = 1.$$

Observation. In case the proper numbers of the matrix are multiple, but to each proper number there correspond as many linearly independent proper vectors as the multiplicity of the root, these indicated properties of the proper vectors hold as before.

§ 4. THE JORDAN CANONICAL FORM

We have proved above that if a matrix has n distinct proper numbers, it may be brought into diagonal form by a similarity transformation. Given the presence of multiple roots, however, such a transformation may not always be possible. Nonetheless the problem of reducing the matrix to as simple a form as possible by a similarity transformation can well be posed. The problem is equivalent to discovering a basis with respect to which the linear transformation connected with the given matrix would have a matrix of simplest form, and the latter proves to be the Jordan canonical form.

Proof of a fundamental theorem to the effect that any matrix may be brought into the Jordan canonical form by a similarity transformation is rather complicated and we will not dwell on it here.[1] We shall limit ourselves to a description of this canonical form.

---

[1] [See FRAZER, R. A., DUNCAN, W. J., AND COLLAR, A. R., [1], § 3.16 for the gist of the proof, or, for the proof proper, TURNBULL, H. W., AND AITKEN, A. C., [1], Chapters V–VI.]

A matrix of the following form is called a *canonical box*:

$$
\begin{bmatrix}
\lambda_i & 0 & 0 & \ldots & 0 & 0 \\
1 & \lambda_i & 0 & \ldots & 0 & 0 \\
0 & 1 & \lambda_i & \ldots & 0 & 0 \\
\cdot & \cdot & \cdot & \cdot & \cdot & \cdot \\
0 & 0 & 0 & \ldots & 1 & \lambda_i
\end{bmatrix}.
$$

On its principal diagonal the single number λ_i is everywhere to be found; directly under the diagonal (in the *subdiagonal*) are disposed elements that are all units; all the rest of the elements are zero.

A canonical box cannot be simplified by utilizing a similarity transformation. It is obvious that a canonical box has the sole multiple proper number λ_i. It may be easily verified that a canonical box has only one proper vector. The minimum polynomial of a box coincides with its characteristic polynomial, viz., it equals $(\lambda_i - \lambda)^{m_i}$ where m_i is the order of the box.

The *Jordan (classical) canonical* form is a quasi-diagonal matrix composed of canonical boxes:

$$
\begin{bmatrix}
\lambda_1 & 0 & 0 & \ldots & 0 & & & & & & \\
1 & \lambda_1 & 0 & \ldots & 0 & & & & & & \\
0 & 1 & \lambda_1 & \ldots & 0 & & & & & & \\
\cdot & \cdot & \cdot & \cdot & \cdot & \cdot & & & & & \\
0 & 0 & 0 & \ldots & \lambda_1 & & & & & & \\
& & & & & \ddots & & & & & \\
& & & & & & \lambda_s & 0 & 0 & \ldots & 0 \\
& & & & & & 1 & \lambda_s & 0 & \ldots & 0 \\
& & & & & & 0 & 1 & \lambda_s & \ldots & 0 \\
& & & & & & \cdot & \cdot & \cdot & \cdot & \cdot \\
& & & & & & 0 & 0 & 0 & \ldots & \lambda_s
\end{bmatrix}
$$

It is admissible that the same number λ_i appear in several canonical boxes.

All the numbers λ_i figuring in the different boxes are proper numbers of the canonical matrix, and the multiplicity of a proper number equals the sum of the orders of the boxes in which it figures as diagonal element. In proof of this, by the theorem concerning the determinant of a quasi-diagonal matrix, the characteristic polynomial of the canonical matrix equals the product of the characteristic polynomials of the separate boxes, each of them equal to $(\lambda_i - \lambda)^{m_i}$, where λ_i is the proper number and m_i is the order of the ith box. Hence follows directly the statement under proof.

The determination of the Jordan boxes for a given matrix A presents certain difficulties. The characteristic polynomial $\Pi(\lambda_i - \lambda)^{m_i}$ coincides with the characteristic polynomial of the original matrix, and it is consequently possible to find it without knowing the canonical matrix itself. Nonetheless, a knowledge of the characteristic polynomial still does not make possible the complete determination of the canonical form, for a proper number λ_i of multiplicity k_i there may correspond several Jordan boxes containing this number as a diagonal element, and regarding them only the sum of their orders will be known, not the order of each box in particular. If the canonical form is to be fully determined, a knowledge of the "elementary divisors" of the matrix must be drawn upon.

Designate by $D_i(\lambda)$ the greatest common divisor of all the minors of the ith order of the determinant $|A - \lambda I|$. In particular, $D_n(\lambda)$ coincides with the characteristic polynomial. It can be proved that all $D_i(\lambda)$, as $D_n(\lambda)$, are common to the class of similar matrices. It can be proved, moreover, that $D_{i-1}(\lambda)$ divides $D_i(\lambda)$[1].

Put

$$\frac{D_i(\lambda)}{D_{i-1}(\lambda)} = E_i(\lambda);$$

obviously

$$D_n(\lambda) = \prod_{i=1}^{n} E_i(\lambda).$$

[1] [See, e.g., TURNBULL, H. W., AND AITKEN, A. C., [1], p. 23 ff.]

It turns out, moreover, that $E_n(\lambda) = \dfrac{D_n(\lambda)}{D_{n-1}(\lambda)}$ is the minimum polynomial of the matrix.

Resolve $E_i(\lambda)$ into linear factors. Then

$$E_i(\lambda) = \prod_{j=1}^{s} (\lambda_j - \lambda)^{m_{ij}}.$$

Here s denotes the number of distinct proper numbers, $\sum\limits_{i=1}^{n} m_{ij} = k_j$; $\sum\limits_{j=1}^{s} \sum\limits_{i=1}^{n} m_{ij} = n$. It is obvious that among the exponents m_{ij} only some will be different from zero.

The binomials $(\lambda_j - \lambda)^{m_{ij}}$ are known as the elementary divisors of the matrix A. A knowledge of the elementary divisors permits us to construct the canonical form, viz.: the Jordan boxes are constructed by starting from the numbers λ_j, and the orders of these boxes are equal to the exponents m_{ij}. The number of boxes containing λ_j equals the number of exponents m_{ij} not equal to zero.

In case the elementary divisors are linear, i.e., if all the non-zero exponents m_{ij} are equal to one, the Jordan boxes degenerate into diagonal elements, and the canonical form turns out to be simply a diagonal form, wherein, of course, a single proper number will appear as often as a diagonal element as it has multiplicity as a zero of the characteristic polynomial.

The converse is also obvious, for it is clear that if a matrix can be brought to diagonal form, its elementary divisors are linear. Therefore *matrices with distinct proper numbers, as also symmetric matrices, have linear elementary divisors.*

If all the elementary divisors $(\lambda_j - \lambda)^{m_{ij}}$ are relatively prime (which occurs only in case $D_{n-1}(\lambda) = 1$), each proper number appears in only one canonical box, and the order of the box equals the multiplicity of the corresponding proper number. Only in this case does the minimum polynomial coincide with the characteristic polynomial.

Let us now consider the matrix transforming a given matrix A into canonical form. With this object in view we introduce into the discussion the linear transformation connected with the matrix A with respect to the initial basis. Then the columns of the transforming matrix will be components of the vectors of that basis in terms of which the linear transformation in question is describable as a canonical matrix.

Let this canonical matrix have the form

$$\begin{bmatrix} \Lambda_1 & & & & \\ & \Lambda_2 & & & \\ & & \cdot & & \\ & & & \cdot & \\ & & & & \Lambda_s \end{bmatrix},$$

where

$$\Lambda_r = \begin{bmatrix} \lambda_r & 0 & \ldots & 0 \\ 1 & \lambda_r & \ldots & 0 \\ \cdot & \cdot & \cdots & \cdot \\ 0 & 0 & \ldots & \lambda_r \end{bmatrix}$$

and the order of Λ_r equals m_r; let $U_1^{(1)}, U_2^{(1)}, \ldots, U_{m_1}^{(1)}, \ldots, U_1^{(s)}, \ldots, U_{m_s}^{(s)}$ be the corresponding basis. Then the following formulas for the transformation hold:

$$AU_1^{(r)} = \lambda_r U_1^{(r)} + U_2^{(r)}$$

$$AU_2^{(r)} = \lambda_r U_2^{(r)} + U_3^{(r)}$$

$$\cdot \quad \cdot \quad \cdot \quad \cdot \quad \cdot \quad \cdot \quad \cdot \quad \cdot$$

$$AU_{m_r-1}^{(r)} = \lambda_r U_{m_r-1}^{(r)} + U_{m_r}^{(r)}$$

$$AU_{m_r}^{(r)} = \lambda_r U_{m_r}^{(r)}$$

for all $r = 1, \ldots, s$.

We see that among the vectors of the canonical basis are the proper vectors of the given matrix, one per box. It can be proved that with this all linearly independent proper vectors of the matrix are exhausted, and consequently the number of linearly independent proper vectors of the given matrix equals the number of canonical boxes in its canonical form. In particular, the number of linearly independent proper vectors belonging to a given proper number equals the number of canonical boxes containing this number.   It is not of greater multiplicity than the proper number, and is equal to this multiplicity in case, and only in case, all boxes containing the given proper number are of order 1, i.e., when the corresponding elementary divisors are linear.

§ 5. THE CONCEPT OF LIMIT FOR VECTORS AND MATRICES

Let a sequence of vectors $X^{(1)}, X^{(2)}, \ldots, X^{(k)}, \ldots$ with components $(x_1^{(1)}, \ldots, x_n^{(1)}), \ldots, (x_1^{(k)}, \ldots, x_n^{(k)}), \ldots$ be given.   If a limit exists for each component: $\lim\limits_{k \to \infty} x_i^{(k)} = x_i$, the vector X, with components $x_1, \ldots, x_n$, is called the *limit* of the sequence $X^{(1)}$, $X^{(2)}, \ldots, X^{(k)}, \ldots$ and the sequence itself is said to be *convergent* to the vector X.   This is written in the form $X^{(k)} \to X$ or $\lim\limits_{k \to \infty} X^{(k)} = X$.

In the same fashion, given a sequence of square matrices $A^{(1)}$, $A^{(2)}, \ldots, A^{(k)}, \ldots$ with elements $(a_{ij}^{(1)}), (a_{ij}^{(2)}), \ldots, (a_{ij}^{(k)}), \ldots$, the matrix A with elements $a_{ij} = \lim\limits_{k \to \infty} a_{ij}^{(k)}$ is called the *limit* of the sequence, if all these limits exist.

In accordance with such a definition of a limit, an *infinite series of vectors* $X^{(1)} + X^{(2)} + \ldots + X^{(k)} + \cdots$ is said to be *convergent* if $\lim\limits_{k \to \infty} (X^{(1)} + X^{(2)} + \cdots + X^{(k)})$ exists; this limit is called the *sum* of the given series.   Obviously it is necessary and sufficient for the convergence of a series of vectors that all the series composed of their corresponding components, i.e., components bearing the same indices, converge; the sums of these series are the components of the sum of the series of vectors.

The concept of the convergence of a series of matrices is defined analogously.

In applied questions it is usually important to judge not only the convergence of a sequence or series, but also to judge the *rate* of this convergence.   With this object in view, the introduction of the norms of vectors and matrices is quite useful.   A norm may be introduced in different ways, and in different cases one or other norm will prove to be most convenient.

Generally, the *norm of a vector* X is an associated non-negative number $\|X\|$ satisfying the following requirements:

1) $\|X\| > 0$ for $X \neq 0$ and $\|0\| = 0$;

2) $\|cX\| = |c|\,\|X\|$ for any numerical multiplier c;

3) $\|X+Y\| \leqslant \|X\|+\|Y\|$ (the "triangular inequality").

From requirements 2) and 3) it is readily deduced that

$$\|X-Y\| \geqslant \Big|\,\|X\|-\|Y\|\,\Big|.$$

Indeed, we have

$$\|X\| = \|X-Y+Y\| \leqslant \|X-Y\|+\|Y\|$$

and therefore

$$\|X-Y\| \geqslant \|X\|-\|Y\|.$$

But

$$\|X-Y\| = \|Y-X\| \geqslant \|Y\|-\|X\|.$$

Consequently

$$\|X-Y\| \geqslant \Big|\,\|X\|-\|Y\|\,\Big|.$$

We shall henceforth make use of the following three ways of assigning a norm: if $X = (x_1, x_2, \ldots, x_n)$,

$$\left\{ \begin{array}{ll} \text{I.} & \|X\|_{\text{I}} = \max_i |x_i| \\[2mm] \text{II.} & \|X\|_{\text{II}} = |x_1|+|x_2|+\cdots+|x_n| \\[2mm] \text{III.} & \|X\|_{\text{III}} = \sqrt{|x_1|^2+|x_2|^2+\cdots+|x_n|^2}. \end{array} \right.$$

It is obvious that for all three norms all the requirements 1)—3) are fulfilled.

The concept of the norm of a vector generalizes the concept of the length of a vector, since for length all the requirements 1)—3) are fulfilled. The third norm introduced by us is indeed none other than the length of the vector.

Furthermore, it is easily established that a necessary and sufficient condition that the sequence of vectors $X^{(k)}$ converge to the vector X is that $\|X^{(k)}-X\| \to 0$ for each of the three norms indicated. For the

first norm this is obvious. For the second and third norm this
follows from the obvious inequalities

$$\|X\|_{\mathrm{I}} \leqslant \|X\|_{\mathrm{II}} \leqslant n\|X\|_{\mathrm{I}}$$
$$\|X\|_{\mathrm{I}} \leqslant \|X\|_{\mathrm{III}} \leqslant \sqrt{n}\|X\|_{\mathrm{I}}.$$

It is easily shown that for convergence of a sequence of vectors
$X^{(k)}$ to a vector X it is necessary and sufficient that $\|X - X^{(k)}\| \to 0$,
whatever norm satisfying conditions 1)—3) we may choose. Here,
if $X^{(k)} \to X$, $\|X^{(k)}\| \to \|X\|$, for $\left| \|X\| - \|X^{(k)}\| \right| \leqslant \|X - X^{(k)}\| \to 0$.

In an analogous fashion, the norm of a square matrix A is a non-
negative number $\|A\|$ satisfying the conditions

$$
\left\{
\begin{array}{llll}
1) & \|A\| & > & 0 \text{ if } A \neq 0 \text{ and } \|0\| = 0; \\
2) & \|cA\| = |c|\ \|A\|; \\
3) & \|A+B\| \leqslant \|A\| + \|B\|; \\
4) & \|AB\| \leqslant \|A\|\ \|B\|.
\end{array}
\right.
$$

Just as in the case of the norms of vectors, the condition $\|A^{(k)} - A\| \to 0$
is necessary and sufficient in order that $A^{(k)} \to A$, and just as in the
case of the norms of vectors, it follows from $A^{(k)} \to A$ that $\|A^{(k)}\| \to \|A\|$.

The norm of a matrix may be introduced in an infinite variety of
ways. Because in the majority of problems connected with estimates both matrices and vectors appear simultaneously in the
reasoning, it is convenient to introduce the norm of a matrix in such
a way that it will be rationally connected with the vector norms
employed in the argument in hand. We shall say that the norm
of a matrix is *compatible* with a given norm of vectors if for any
matrix A and any vector X the following inequality is satisfied:

$$\|AX\| \leqslant \|A\|\ \|X\|.$$

We will now indicate a device making it possible to construct the
matrix norm so as to render it compatible with a given vector norm,
to wit: we shall adopt for the norm of the matrix A the maximum
of the norms of the vectors AX on the assumption that the vector X
runs over the set of all vectors whose norm equals unity:

$$\|A\| = \max_{\|X\|=1} \|AX\|.$$

In consequence of the continuity of a norm, for each matrix A this maximum is attainable, i.e., a vector X_0 can be found such that $\|X_0\| = 1$ and $\|AX_0\| = \|A\|$.

We shall prove that a norm constructed in such a manner satisfies requirements 1)—4), set previously, and the compatibility condition.

Let us begin with the verification of the *first requirement*.

Let $A \neq 0$. Then a vector X, $\|X\| = 1$, can be found such that $AX \neq 0$, and accordingly $\|AX\| \neq 0$. Therefore $\|A\| = \max_{\|X\|=1} \|AX\| \neq 0$.

If, however, $A = 0$, $\|A\| = \max_{\|X\|=1} \|0X\| = 0$.

Second requirement. On the strength of the definition, $\|cA\| = \max \|cAX\|$. Obviously $\|cAX\| = |c| \, \|AX\|$ and thus

$$\|cA\| = \max_{\|X\|=1} |c| \, \|AX\| = |c| \max \|AX\| = |c| \, \|A\|.$$

Let us verify, furthermore, the compatibility condition.

Let $Y \neq 0$ be any vector; then $X = \dfrac{1}{\|Y\|} Y$ will satisfy the condition that $\|X\| = 1$. Consequently

$$\|AY\| = \|A(\|Y\|X)\| = \|Y\| \, \|AX\| \leqslant \|Y\| \, \|A\|.$$

Third requirement. For the matrix $A + B$ find a vector X_0 such that $\|A + B\| = \|(A + B)X_0\|$ and $\|X_0\| = 1$. Then

$$\|A + B\| = \|(A + B)X_0\|$$
$$= \|AX_0 + BX_0\| \leqslant \|AX_0\| + \|BX_0\| \leqslant \|A\| \, \|X_0\| + \|B\| \, \|X_0\|$$
$$= \|A\| + \|B\|.$$

Lastly, the *fourth requirement.* For the matrix AB find a vector X_0 such that $\|X_0\| = 1$ and $\|ABX_0\| = \|AB\|$. Then

$$\|AB\| = \|ABX_0\|$$
$$= \|A(BX_0)\| \leqslant \|A\| \, \|BX_0\| \leqslant \|A\| \, \|B\| \, \|X_0\|$$
$$= \|A\| \, \|B\|.$$

We have verified the satisfaction of all four requirements and the compatibility condition. A matrix norm constructed in this manner we shall speak of as *subordinate* to the given norm of vectors. It is obvious that for any matrix norm, subordinate to whatsoever vector norm, $\|I\| = 1$.

Let us now construct matrix norms subordinate to the three norms of vectors introduced above.

I. $$\|X\|_{\mathrm{I}} = \max_i |x_i|.$$

The matrix norm subordinate to this vector norm is

$$\|A\|_{\mathrm{I}} = \max_i \sum_{k=1}^n |a_{ik}|.$$

In proof, let $\|X\| = 1$. Then

$$\|AX\| = \max_i \Big| \sum_{k=1}^n a_{ik} x_k \Big| \leqslant \max_i \sum_{k=1}^n |a_{ik}| \, |x_k| \leqslant \max_i \sum_{k=1}^n |a_{ik}|.$$

Consequently

$$\max_{\|X\|=1} \|AX\| \leqslant \max_i \sum_{k=1}^n |a_{ik}|.$$

We shall now prove that $\max\limits_{\|X\|=1} \|AX\|$ is in fact equal to $\max\limits_i \sum\limits_{k=1}^n |a_{ik}|$. For this we shall construct a vector X_0 such that $\|X_0\| = 1$ and $\|AX_0\| = \max\limits_i \sum\limits_{k=1}^n |a_{ik}|$. Letting $\sum\limits_{k=1}^n |a_{ik}|$ attain its greatest value for $i=j$, and then taking as the component $x_k^{(0)}$ of the vector X_0: $x_k^{(0)} = \dfrac{|a_{jk}|}{a_{jk}}$, if $a_{jk} \neq 0$, and $x_k^{(0)} = 1$ if $a_{jk} = 0$, we have, obviously, $\|X_0\| = 1$. Furthermore,

$$\Big| \sum_{k=1}^n a_{ik} x_k^{(0)} \Big| \leqslant \sum_{k=1}^n |a_{ik}| \leqslant \sum_{k=1}^n |a_{jk}| \quad \text{for } i \neq j$$

and

$$\Big| \sum_{k=1}^n a_{jk} x_k^{(0)} \Big| = \sum_{k=1}^n |a_{jk}|.$$

Consequently

$$\max_i \Big| \sum_{k=1}^n a_{ik} x_k^{(0)} \Big| = \sum_{k=1}^n |a_{jk}| = \max_i \sum_{k=1}^n |a_{ik}|.$$

Thus $\|AX_0\| = \max\limits_{i} \sum\limits_{k=1}^{n} |a_{ik}|$, Q.E.D.

II.
$$\|X\|_{\mathrm{II}} = \sum_{i=1}^{n} |x_i|.$$

The matrix norm subordinate to this vector norm is

$$\|A\|_{\mathrm{II}} = \max_{k} \sum_{i=1}^{n} |a_{ik}|.$$

In proof thereof, let $\|X\| = 1$, then

$$\|AX\| = \sum_{i=1}^{n} \left| \sum_{k=1}^{n} a_{ik}x_k \right| \leqslant \sum_{i=1}^{n} \sum_{k=1}^{n} |a_{ik}|\,|x_k|$$

$$\leqslant \sum_{k=1}^{n} |x_k| \left[\sum_{i=1}^{n} |a_{ik}| \right]$$

$$\leqslant \left[\max_{k} \sum_{i=1}^{n} |a_{ik}| \right] \sum_{k=1}^{n} |x_k| \leqslant \max_{k} \sum_{i=1}^{n} |a_{ik}|.$$

Now let us take a vector X_0 of the following form: let $\sum\limits_{i=1}^{n} |a_{ik}|$ attain its greatest value for the column numbered j. Put $x_k^{(0)} = 0$ for $k \neq j$ and $x_j^{(0)} = 1$. Obviously a vector constructed in this manner has its norm equal to unity. Furthermore

$$\|AX_0\| = \sum_{i=1}^{n} \left| \sum_{k=1}^{n} a_{ik}x_k^{(0)} \right| = \sum_{i=1}^{n} |a_{ij}| = \max_{k} \sum_{k=1}^{n} |a_{ik}|.$$

Thus

$$\max \|AX_0\| = \max_{k} \sum_{i=1}^{n} |a_{ik}|,$$

Q.E.D.

III.
$$\|X\|_{\mathrm{III}}^2 = \sum_{k=1}^{n} |x_k|^2 = (X, X).$$

The matrix norm subordinate to this vector norm is

$$\|A\|_{\mathrm{III}} = \sqrt{\lambda_1},$$

where λ_1 is the largest proper number of the matrix $A'A$. In proof, we have

$$\|A\| = \max_{\|X\|=1} \|AX\|;$$

but

$$\|AX\|_{\mathrm{III}}^2 = (AX, AX) = (X, A'AX).$$

The matrix $A'A$ is symmetric.   Let $\lambda_1 \geqslant \lambda_2 \geqslant \cdots \geqslant \lambda_n$ be its proper numbers and $X_1, X_2, \ldots, X_n$ be the orthonormal system of proper vectors belonging to these proper numbers.

Now take any vector X with its norm equal to unity and resolve it in terms of the proper vectors:

$$X = c_1 X_1 + c_2 X_2 + \cdots + c_n X_n.$$

Then

$$(X, X) = c_1^2 + c_2^2 + \cdots + c_n^2 = 1.$$

Moreover,

$$\begin{aligned}
\|AX\|^2 &= (X, A'AX) \\
&= (c_1 X_1 + \cdots + c_n X_n, c_1 \lambda_1 X_1 + \cdots + c_n \lambda_n X_n) \\
&= \lambda_1 c_1^2 + \cdots + \lambda_n c_n^2 \leqslant \lambda_1 (c_1^2 + \cdots + c_n^2) = \lambda_1.
\end{aligned}$$

For the vector $X = X_1$:

$$\|AX_1\|^2 = (X_1, A'AX_1) = (X_1, \lambda_1 X_1) = \lambda_1.$$

Thus

$$\max_{\|X\|=1} \|AX\| = \sqrt{\lambda_1},$$

Q.E.D.

We shall now prove several theorems connected with the concept of limit.

THEOREM 1.   *In order that $A^m \rightarrow 0$, it is necessary and sufficient that all the proper numbers of the matrix A be of modulus less than unity.*

Proof.   Assume for simplicity that the matrix A can be brought into diagonal form: $A = C\Lambda C^{-1}$, where $\Lambda = \ulcorner \lambda_1, \lambda_2, \ldots, \lambda_n \lrcorner$ and $\lambda_1, \lambda_2, \ldots, \lambda_n$ are the proper numbers of matrix A.   Then $A^m = C\Lambda^m C^{-1}$.   It is obvious that $\Lambda^m = \ulcorner \lambda_1^m, \lambda_2^m, \ldots, \lambda_n^m \lrcorner$.   In order that $A^m \rightarrow 0$, it is necessary and sufficient that $\Lambda^m \rightarrow 0$, for which it is in turn necessary and sufficient that all proper numbers $\lambda_1, \lambda_2, \ldots, \lambda_n$ have modulus less than unity.

In case the matrix A cannot be brought into diagonal form, the theorem is proved either with the aid of considerations of continuity or by passing to the Jordan canonical form. We shall not dwell on the details of this proof.

The conditions given in Theorem 1 are inconvenient for testing, inasmuch as they require foreknowledge of the proper numbers of the matrix A. We shall therefore establish some simpler sufficient conditions rendering $\lim\limits_{m \to \infty} A^m = 0$.

THEOREM 2. *In order that $A^m \to 0$, it is sufficient that any one of the norms of A be less than unity.*

Proof. On the strength of the fourth requirement of a norm, we have

$$\|A^m\| \leqslant \|A^{m-1}\|\,\|A\| \leqslant \|A^{m-2}\|\,\|A\|^2 \leqslant \cdots \leqslant \|A\|^m.$$

Therefore $\|A^m\| \to 0$ if $\|A\| < 1$, and thus, in view of the foregoing, $A^m \to 0$.

Combining Theorems 1 and 2, we arrive at the following result:

THEOREM 3. *No proper number of a matrix exceeds any of its norms in modulus.*

Proof. Let $\|A\| = a$. Consider a matrix $B = \dfrac{1}{a + \varepsilon}\,A$, where ε is any positive number. We have

$$\|B\| = \frac{a}{a + \varepsilon} < 1,$$

and accordingly $B^m \to 0$ as $m \to \infty$. On the strength of Theorem 1 its proper numbers have modulus less than unity. But it is obvious that the proper numbers of the matrix B equal $\dfrac{1}{a + \varepsilon}\,\lambda_i$, where λ_i are the proper numbers of the matrix A. Thus $\dfrac{|\lambda_i|}{a + \varepsilon} < 1$, i.e., $|\lambda_i| < a + \varepsilon$. Since ε may be taken arbitrarily small, $|\lambda_i| \leqslant a$.

THEOREM 4. *In order that the series*

(1) $$I + A + \cdots + A^m + \cdots$$

converge, it is necessary and sufficient that $A^m \to 0$ as $m \to \infty$. In such a case the sum of series (1) equals $(I - A)^{-1}$.

Proof. The necessity of this condition is obvious. We shall show that it is sufficient.

On the strength of Theorem 1, all proper numbers of the matrix A are less than of unit modulus.

Accordingly

$$|I - A| \neq 0$$

and therefore $(I - A)^{-1}$ exists.

Consider the identity

$$(I + A + A^2 + \cdots + A^k)(I - A) = I - A^{k+1}.$$

Postmultiplying it by $(I - A)^{-1}$, we obtain

$$I + A + A^2 + \cdots + A^k = (I - A)^{-1} - A^{k+1}(I - A)^{-1},$$

whence it follows that, as $k \to \infty$,

$$I + A + \cdots + A^k \to (I - A)^{-1}$$

since $A^{k+1} \to 0$.

Thus

$$I + A + \cdots + A^k + \cdots = (I - A)^{-1},$$

which is what was required to be proved.

In the light of Theorem 1, the necessary and sufficient condition for the convergence of the series (1) is the inequality $|\lambda_i| < 1$ for all proper numbers of the matrix A. A sufficient criterion of convergence, in view of Theorem 2, is the inequality $\|A\| < 1$, whatever one of the norms be employed. Given that this condition is satisfied, it is easy to give the following estimate of the rate of convergence of the series (1):

THEOREM 5. *If* $\|A\| < 1$,

$$\|(I - A)^{-1} - (I + A + \cdots + A^k)\| \leqslant \frac{\|A\|^{k+1}}{1 - \|A\|}.$$

Proof. We have:

$$(I - A)^{-1} - (I + A + \cdots + A^k) = A^{k+1} + A^{k+2} + \cdots$$

whence

$$\|(I - A)^{-1} - (I + A + \cdots + A^k)\| \leqslant \|A\|^{k+1} + \|A\|^{k+2} + \cdots = \frac{\|A\|^{k+1}}{1 - \|A\|}$$

and the theorem is proved.

CHAPTER II

SYSTEMS OF LINEAR EQUATIONS

This chapter is devoted to three problems that are closely related to each other: the problem of solving a non-homogeneous linear algebraic system, the problem of inverting a matrix, and the problem spoken of as elimination.

In theory all of these problems are soluble simply enough. However, if a matrix of high order is connected with the problem, the actual solution requires a great number of computational operations.

Numerical methods solving the stated problems are divisible into two groups: *exact* and *iterative methods*. By *exact* methods we understand methods that give the solution of a problem by means of a finite number of elementary arithmetic operations. The number of computational operations necessary for the solution of the problem depends only upon the form of the computational scheme and upon the order of the matrix defining the given problem. Inexactitude in the solution found occurs as the result of the inevitable rounding-off of the figures in the course of the computation. Along with this, one may run up against the phenomenon of the disappearance of significant figures in the course of the computation, as the result of the subtraction of two numbers differing little from each other. This loss of significant figures may occasion such an important reduction in the accuracy of the result that it is often necessary to alter the computational scheme because of it, or re-do the work with a greater number of significant figures in the intermediate calculations.

The fundamental method of this group is the method based on the idea of *elimination*. The algorithm of this method, with which the name of Gauss is associated, consists—when applied to the solution of a non-homogeneous linear system—of a chain of successive eliminations by means of which the given system is transformed into a system with a triangular matrix whose solution presents no difficulty.

Many different computational schemes have been developed and are currently employed for all three of the problems mentioned.

Those schemes known as *compact schemes* are outstanding among these. The compact schemes are based on the possibility of conducting the calculations of expressions of the form $\dfrac{\sum\limits_{k=1}^{n} a_k b_k}{c}$ all at one time on the modern calculating machine, dispensing with the recording of the results of the intermediate computations of the separate summands $a_k b_k$; this type of computation is spoken of as *accumulation*. By accumulative computation the rounding error is naturally reduced sharply and the volume of intermediate results to be recorded is considerably diminished.

A method founded on the idea of *bordering* also belongs to the exact methods. At the root of this method lies the representation of a given matrix in the form of matrices situated one within the next, starting with a matrix of the first order.

Iterative methods afford a means by which a system of linear equations may be solved approximately. The solution of a system by iterative methods is obtained as the limit of successive approximations computed by some uniform process. The convergence of these approximations depends essentially on the elements of the matrix defining the given problem. The rate of the convergence depends also on a happy choice of the initial approximation on which the iterative process is founded.

In particular, for the basic iterative processes, there exist matrices for which a process converges only slowly, or even diverges.

The immense advantage of the iterative schemes consists in the simplicity and uniformity of the operations to be effected, and therefore in the possibility of completely mechanizing the process of computation.

In the continuation of this chapter, we shall assume that all the given numbers defining the condition of the problem are *real*.

§ 6. GAUSS'S METHOD

In this section we shall recall to the reader the well-known algorithm for solving a system of equations by a successive elimination of the unknowns. As has already been mentioned above, it is associated with the name of Gauss. For simplicity of reasoning and in order to carry the calculation to the end, we shall consider a system of four equations in four unknowns.

Thus let there be given the system of equations

$$
\begin{aligned}
a_{11}x_1 + a_{12}x_2 + a_{13}x_3 + a_{14}x_4 &= a_{15} \\
a_{21}x_1 + a_{22}x_2 + a_{23}x_3 + a_{24}x_4 &= a_{25} \\
a_{31}x_1 + a_{32}x_2 + a_{33}x_3 + a_{34}x_4 &= a_{35} \\
a_{41}x_1 + a_{42}x_2 + a_{43}x_3 + a_{44}x_4 &= a_{45}
\end{aligned}
$$

(1)

Let us divide the coefficients of the first of equations (1) by the coefficient a_{11}, which we shall speak of as the *leading* element, and use the designation

$$
(2) \qquad b_{1j} = \frac{a_{1j}}{a_{11}}; \quad j > 1.
$$

We will have obtained a new equation

$$
(3) \qquad x_1 + b_{12}x_2 + b_{13}x_3 + b_{14}x_4 = b_{15}.
$$

Let us now pass on to an auxiliary system, composed of the three equations in three unknowns that are obtained as the result of eliminating x_1 from the last three equations of (1) by means of equation (3). This elimination is easily done by multiplying equation (3) by a_{21}, a_{31}, a_{41} in turn (i.e., by the leading elements of the second, third and fourth rows) and subtracting it from the corresponding equations of system (1). The coefficients of the new system obtained as the result of the elimination of one variable, we shall designate by $a_{ij \cdot 1}$:

$$
(4) \qquad a_{ij \cdot 1} = a_{ij} - a_{i1}b_{1j}; \quad i, j \geqslant 2.
$$

Proceeding, we shall now divide the coefficients of the first equation of the new system by the "leading" element $a_{22 \cdot 1}$.

We shall have obtained the equation

(5)
$$x_2 + b_{23 \cdot 1}x_3 + b_{24 \cdot 1}x_4 = b_{25 \cdot 1},$$

where

(6)
$$b_{2j \cdot 1} = \frac{a_{2j \cdot 1}}{a_{22 \cdot 1}}; \quad j > 2.$$

Continuing the process analogously, we shall pass to a system of two equations in two unknowns, whose coefficients have the form

(7)
$$a_{ij \cdot 2} = a_{ij \cdot 1} - a_{i2 \cdot 1}b_{2j \cdot 1}, \quad i, j \geqslant 3.$$

Dividing the coefficients of the first equation of this system by the leading element $a_{33 \cdot 2}$ and defining

(8)
$$b_{3j \cdot 2} = \frac{a_{3j \cdot 2}}{a_{33 \cdot 2}}; \quad j > 3,$$

we shall obtain the equation

(9)
$$x_3 + b_{34 \cdot 2}x_4 = b_{35 \cdot 2}.$$

Finally, a last step will lead us to a single equation containing a single unknown with coefficient $a_{44 \cdot 3}$. Upon dividing this equation by $a_{44 \cdot 3}$, we obtain

$$x_4 = b_{45 \cdot 3}.$$

Collecting all equations with coefficients $b_{ij \cdot i-1}, j > i$, we shall have obtained a triangular system equivalent to the given one; its solution will be the solution of our system. We note that the process described is possible only on condition that none of the leading elements by which we have, in the course of the work, divided the coefficients of the first equations of the intermediate systems, are equal to zero.

So for solving the given system we first construct an auxiliary triangular system, and then solve it. The process of finding the

TABLE I. Single-Division Scheme

x_1	x_2	x_3	x_4	Σ	Σ						Σ
a_{11}	a_{12}	a_{13}	a_{14}	a_{15}	a_{16}	1.00	0.42	0.54	0.66	0.3	2.92
a_{21}	a_{22}	a_{23}	a_{24}	a_{25}	a_{26}	0.42	1.00	0.32	0.44	0.5	2.68
a_{31}	a_{32}	a_{33}	a_{34}	a_{35}	a_{36}	0.54	0.32	1.00	0.22	0.7	2.78
a_{41}	a_{42}	a_{43}	a_{44}	a_{45}	a_{46}	0.66	0.44	0.22	1.00	0.9	3.22
1	b_{12}	b_{13}	b_{14}	b_{15}	b_{16}	1	0.42	0.54	0.66	0.3	2.92
	$a_{22\cdot1}$	$a_{23\cdot1}$	$a_{24\cdot1}$	$a_{25\cdot1}$	$a_{26\cdot1}$		0.82360	0.09320	0.16280	0.37400	1.45360
	$a_{32\cdot1}$	$a_{33\cdot1}$	$a_{34\cdot1}$	$a_{35\cdot1}$	$a_{36\cdot1}$		0.09320	0.70840	−0.13640	0.53800	1.20320
	$a_{42\cdot1}$	$a_{43\cdot1}$	$a_{44\cdot1}$	$a_{45\cdot1}$	$a_{46\cdot1}$		0.16280	−0.13640	0.56440	0.70200	1.29280
	1	$b_{23\cdot1}$	$b_{24\cdot1}$	$b_{25\cdot1}$	$b_{26\cdot1}$		1	0.11316	0.19767	0.45410	1.76493
		$a_{33\cdot2}$	$a_{34\cdot2}$	$a_{35\cdot2}$	$a_{36\cdot2}$			0.69785	−0.15482	0.49568	1.03871
		$a_{43\cdot2}$	$a_{44\cdot2}$	$a_{45\cdot2}$	$a_{46\cdot2}$			−0.15482	0.53222	0.62807	1.00547 (5)
		1	$b_{34\cdot2}$	$b_{35\cdot2}$	$b_{36\cdot2}$			1	−0.22185	0.71030	1.48844
			$a_{44\cdot3}$	$a_{45\cdot3}$	$a_{46\cdot3}$				0.49787	0.73804	1.23591
			1	x_4	$\bar{x}_4$				1	1.48240	2.48240
		1		x_3	$\bar{x}_3$			1		1.03917	2.03916 (7)
	1			x_2	$\bar{x}_2$		1			0.04348	1.04348 (80)
1				x_1	$\bar{x}_1$	1				−1.25780	−0.25779

Horizontal sums in superior position.

coefficients of the triangular system we shall call the *forward* course, and the process of obtaining its solution, the *return* course, or *back solution*.

The scheme that we have described we shall call *the single-division scheme* (see Table I).

We shall say a few words concerning a control (check) that it is expedient to employ when computing by the single-division scheme. This control is based on the following circumstance. If in the system given we make the substitution $\bar{x}_i = x_i + 1$, then for determining the $\bar{x}_i$ we will obtain a system with the former coefficients, and with its constant side equal to the sums of the elements of the rows of the matrix of the coefficients (plus the constant terms). Therefore having formed the sums of the elements of each row (the control sums), we shall perform upon them the very same operations as upon the rest of the elements of the row. In the absence of computational blunders, the numbers found must coincide with the analogous sums of the transformed rows. The return course is controlled by finding the numbers $\bar{x}_i$ and their coincidence with the numbers $x_i + 1$.

We will briefly explain Table I.

The forward course is carried through as follows. Having copied the matrix of the coefficients (including the constant terms and the control sums), we divide the first row by the leading element and copy the result as the last row of the matrix. Next we compute the elements $a_{ij} \cdot 1$; $i, j \geqslant 2$, of the first auxiliary matrix: taking any element of the given matrix, we subtract from it the product of the leading element of the row to which it belongs by the last element of the column to which it belongs. Continuing, the process is repeated. The computation of the forward course is concluded when we arrive at a matrix consisting of one row.

In the return course, one utilizes the rows containing units, beginning with the last: in the last of these rows, in the column of constant terms, we obtain the value of the last unknown, and in the control column, the control value. Continuing, the values of the unknowns are obtained sequentially, as the result of subtracting from the elements of the next-to-the-last column the products of the corresponding b-coefficients by the values of the unknowns found previously. The unit symbols displayed at the end of the scheme aid in

TABLE II. The Single-Division Scheme: Symmetric Case

x_1	x_2	x_3	x_4		Σ	x_1	x_2	x_3	x_4		Σ
a_{11}	a_{12}	a_{13}	a_{14}	a_{15}	a_{16}	1.00	0.42	0.54	0.66	0.3	2.92
	a_{22}	a_{23}	a_{24}	a_{25}	a_{26}		1.00	0.32	0.44	0.5	2.68
		a_{33}	a_{34}	a_{35}	a_{36}			1.00	0.22	0.7	2.78
			a_{44}	a_{45}	a_{46}				1.00	0.9	3.22
1	b_{12}	b_{13}	b_{14}	b_{15}	b_{16}	1	0.42	0.54	0.66	0.3	2.92
	$a_{22.1}$	$a_{23.1}$	$a_{24.1}$	$a_{25.1}$	$a_{26.1}$		0.82360	0.09320	0.16280	0.37400	1.45360
		$a_{33.1}$	$a_{34.1}$	$a_{35.1}$	$a_{36.1}$			0.70840	−0.13640	0.53800	1.20320
			$a_{44.1}$	$a_{45.1}$	$a_{46.1}$				0.56440	0.70200	1.29280
	1	$b_{23.1}$	$b_{24.1}$	$b_{25.1}$	$b_{26.1}$		1	0.11316	0.19767	0.45410	1.76493
		$a_{33.2}$	$a_{34.2}$	$a_{35.2}$	$a_{36.2}$			0.69785	−0.15482	0.49568	1.03871
			$a_{44.2}$	$a_{45.2}$	$a_{46.2}$				0.53222	0.62807	1.00547
		1	$b_{34.2}$	$b_{35.2}$	$b_{36.2}$			1	−0.22185	0.71030	1.48844
			$a_{44.3}$	$a_{45.3}$	$a_{46.3}$				0.49787	0.73804	1.23591
			1	x_4	$\bar{x}_4$				1	1.48240	2.48240
		1		x_3	$\bar{x}_3$			1		1.03917	2.03916
	1			x_2	$\bar{x}_2$		1			0.04348	1.04348
1				x_1	$\bar{x}_1$	1				−1.25780	−0.25779

locating the coefficients that correspond to the given x in the necessary rows. Thus

(9b) $\quad x_2 = b_{25\cdot1} - b_{24\cdot1}x_4 - b_{23\cdot1}x_3$

$\qquad = 0.45410 - 0.19767 \cdot 1.48240 - 0.11316 \cdot 1.03917 = 0.04348.$

In conclusion we note that the operation of subtraction may be replaced by addition if in copying a row with coefficients b, one reverses their signs.

The number of multiplications and divisions necessary for finding the solution of a system of n equations by the single-division scheme is equal to $\frac{n}{3}(n^2 + 3n - 1)$.

In case the matrix of the coefficients of the system is symmetric, i.e., $a_{ij} = a_{ji}$, we have, obviously, $a_{ij\cdot k} = a_{ji\cdot k}$. We may therefore omit writing the elements below the principal diagonal. The single-division scheme, as modified for symmetric matrix, is shown in Table II.

The leading element of the row, omitted in writing the coefficients, (and which we need for the computation of the elements of the auxiliary matrices) is easily found as the uppermost element of the column containing the diagonal element of the given row. The control column contains, as before, the sums of all the elements of each row, even including those omitted in tabulation.

In case several equations with the same matrix must be solved, their solutions are naturally sought simultaneously, the constant terms being entered in adjacent columns. The control sum is formed as the sum of the elements of the rows of the expanded matrix. The scheme for solution of several equations is given in Table III.

Other computational schemes as well are possible for Gauss's method. We mention a scheme of multiplication and subtraction in which the coefficients of the auxiliary systems, which we shall designate by $A_{ij\cdot k}$, $i, j > k$, are obtained as the difference of the cross-products

$$A_{ij\cdot k} = A_{kk\cdot k-1}A_{ij\cdot k-1} - A_{ik\cdot k-1} \cdot A_{kj\cdot k-1}.$$

With this scheme the forward course requires no divisions at all; they are all moved over into the return course.

TABLE III. *Single-Division Scheme: Several Equations*

x_1	x_2	x_3	x_4				Σ
1.00	0.42	0.54	0.66	0.25	0.3	0.15	3.32
0.42	1.00	0.32	0.44	0.45	0.5	0.30	3.43
0.54	0.32	1.00	0.22	0.65	0.7	0.45	3.88
0.66	0.44	0.22	1.00	0.85	0.9	0.60	4.67
1	0.42	0.54	0.66	0.25	0.3	0.15	3.32
	0.82360	0.09320	0.16280	0.34500	0.37400	0.23700	2.03560
	0.09320	0.70840	-0.13640	0.51500	0.53800	0.36900	2.08720
	0.16280	-0.13640	0.56440	0.68500	0.70200	0.50100	2.47880 (8)
	1	0.11316	0.19767	0.41889	0.45410	0.28776	2.47159
		0.69785	-0.15482	0.47596	0.49568	0.34218	1.85685 (2)
		-0.15482	0.53222	0.61680	0.62807	0.45415	2.07643 (2)
		1	-0.22185	0.68204	0.71030	0.49033	2.66082
			0.49787	0.72239	0.73804	0.53006	2.48838 (6)
			1	1.45096	1.48240	1.06466	4.99805 (2)
				1.00394	1.03917	0.72652	3.76964 (3)
				0.01847	0.04348	-0.00490	1.05705 (6)
1				-1.25752	-1.25780	-0.94294	-2.45828 (6)

Horizontal sums in superior position.

The method of *pivotal condensation* (*dominant elements*) is quite generally encountered. In this scheme the elimination is conducted as follows. Select the largest element of the matrix A, a_{kj}, which is called the pivot (dominant element), and compute the multiplier

$$m_i = -\frac{a_{ij}}{a_{kj}}$$

for all $i \neq k$. Now multiply the pivot row (i.e., the kth row) of the matrix A by the multipliers m_i in turn, and add the result to the corresponding ith row. As a result one obtains a new matrix of which the jth column consists of zeros. Discarding this column and the kth row, we go over to the first auxiliary matrix with a number of rows less by one. The subsequent eliminations are conducted each time by means of the pivotal element of the transformed matrix. The sought-for triangular system is obtained by collecting the pivot rows.

In concluding this section we remark that in computing the coefficients of the intermediate systems by the single-division scheme we may encounter a disappearance of significant figures. If this occurs for a leading element this implies an important loss of accuracy.

In such a case the pivotal condensation scheme may be conveniently used to carry the computations through.

In such modification, the process is always applicable, provided the determinant of the system is different from zero.

§ 7. THE EVALUATION OF DETERMINANTS

Gauss's method, developed in the preceding section for the solution of a linear system, also may be applied to the computation of determinants. We shall dwell separately on a description of the corresponding scheme, since the computation of determinants is frequently met with in applications. Let

$$\Delta = \begin{vmatrix} a_{11} & a_{12} & \cdots & a_{1n} \\ a_{21} & a_{22} & \cdots & a_{2n} \\ \cdot & \cdot & \cdots & \cdot \\ a_{n1} & a_{n2} & \cdots & a_{nn} \end{vmatrix},$$

and let $a_{11} \neq 0$. Remove the element a_{11} from the first row. We then have, using the symbols of § 6,

$$\Delta = a_{11} \begin{vmatrix} 1 & b_{12} & \cdots & b_{1n} \\ a_{21} & a_{22} & \cdots & a_{2n} \\ \vdots & & & \\ a_{n1} & a_{n2} & \cdots & a_{nn} \end{vmatrix}.$$

Next take from each row the first row multiplied by the first element of the particular row in hand. We will obviously obtain

$$\Delta = a_{11} \begin{vmatrix} 1 & b_{12} & \cdots & b_{1n} \\ 0 & a_{22\cdot1} & \cdots & a_{2n\cdot1} \\ \vdots & & & \\ 0 & a_{n2\cdot1} & \cdots & a_{nn\cdot1} \end{vmatrix} = a_{11} \begin{vmatrix} a_{22\cdot1} & \cdots & a_{2n\cdot1} \\ \vdots & & \\ a_{n2\cdot1} & \cdots & a_{nn\cdot1} \end{vmatrix}.$$

With the determinant of the $(n-1)$th order that remains, we shall now deal just as before, provided $a_{22\cdot1} \neq 0$.

Carrying through the process, we shall find that the sought-for determinant is equal to the product of the leading elements:

$$\Delta = a_{11}a_{22\cdot1} \ldots a_{nn\cdot n-1}.$$

If at any step it should turn out that $a_{ii\cdot i-1} = 0$ or $a_{ii\cdot i-1}$ is close to zero (which would imply the reduction of the accuracy of the computations), the order of the rows and columns of the determinant could be rearranged beforehand so that a non-vanishing element appears in the upper left corner.

The best result—in the sense of reliability—will be obtained if at each step of the process the maximum element (in point of absolute value) of the matrix whose determinant is being computed is brought to the upper left corner.

The number of multiplications and divisions necessary for the computation of a determinant of the nth order is equal to $\frac{n-1}{3}(n^2+n+3)$.

In Table IV is given the scheme for computing the determinant of the example.

TABLE IV. *Computation of a Determinant*

				Σ					Σ
a_{11}	a_{12}	a_{13}	a_{14}	a_{15}	1.00	0.42	0.54	0.66	2.62
a_{21}	a_{22}	a_{23}	a_{24}	a_{25}	0.42	1.00	0.32	0.44	2.18
a_{31}	a_{32}	a_{33}	a_{34}	a_{35}	0.54	0.32	1.00	0.22	2.08
a_{41}	a_{42}	a_{43}	a_{44}	a_{45}	0.66	0.44	0.22	1.00	2.32
1	b_{12}	b_{13}	b_{14}	b_{15}	1	0.42	0.54	0.66	2.62
	$a_{22 \cdot 1}$	$a_{23 \cdot 1}$	$a_{24 \cdot 1}$	$a_{25 \cdot 1}$		0.82360	0.09320	0.16280	1.07960
	$a_{32 \cdot 1}$	$a_{33 \cdot 1}$	$a_{34 \cdot 1}$	$a_{35 \cdot 1}$		0.09320	0.70840	−0.13640	0.66520
	$a_{42 \cdot 1}$	$a_{43 \cdot 1}$	$a_{44 \cdot 1}$	$a_{45 \cdot 1}$		0.16280	−0.13640	0.56440	0.59080
	1	$b_{23 \cdot 1}$	$b_{24 \cdot 1}$	$b_{25 \cdot 1}$		1	0.11316	0.19767	1.31083
		$a_{33 \cdot 2}$	$a_{34 \cdot 2}$	$a_{35 \cdot 2}$			0.69785	−0.15482	0.54303
		$a_{43 \cdot 2}$	$a_{44 \cdot 2}$	$a_{45 \cdot 2}$			−0.15482	0.53222	0.37740
		1	$b_{34 \cdot 2}$	$b_{35 \cdot 2}$			1	−0.22185	0.77815
			$a_{44 \cdot 3}$	$a_{45 \cdot 3}$				0.49787	0.49787
					1	0.82360	0.69785	0.49787	0.28615

From an inspection of the process of computing the determinant we see that, with the exception of the last multiplication, it coincides with the forward course of the Gauss process, applied to a system with the matrix for which the determinant is computed.

The familiar rule of Cramer (§ 1) shows that the solution of a linear system may be found in the form $x_i = \dfrac{\Delta_i}{\Delta}$, $i = 1, \ldots, n$, where Δ denotes the determinant of the coefficients of the system, and Δ_i the determinant of the coefficients in which the ith column has been replaced by the constant terms. Thus for the solution of the system one must compute $n+1$ determinants of the nth order.

Comparing the Gauss process for the solution of a system with that for computing a determinant, we see that the volume of computations for the solution of a system only slightly exceeds that for a single determinant. The use of Cramer's formula for the numerical solution of a system is thus inefficient. Essentially in the Gauss method the computation of all the determinants Δ and Δ_i is effected simultaneously, the division by $\Delta = a_{11}a_{22\cdot 1} \ldots a_{nn\cdot n-1}$ being accomplished gradually, one factor per step.

§ 8. COMPACT ARRANGEMENTS FOR THE SOLUTION OF NON-HOMOGENEOUS LINEAR SYSTEMS

In § 6 we saw that the solution of a system of linear equations by the single-division scheme reduced to the determination of the coefficients $a_{ij\cdot k}$ of auxiliary systems (including the constant terms), and the coefficients $b_{ij\cdot i-1}$ of the equations of a final triangular system. Only the coefficients $b_{ij\cdot i-1}$ were necessary to us there for obtaining the solution of the given system, and the numbers $a_{ij\cdot k}$ played an auxiliary role, being necessary only to determine the numbers $b_{ij\cdot i-1}$, which we shall indeed henceforth denote simply by b_{ij}, dropping the index denoting the number of the step.

We shall show[1] that the numbers b_{ij} can be obtained by a process of accumulation permitting one to dispense with the computation and recording of all the coefficients $a_{ij\cdot k}$.

[1] P. Dwyer, [1].

Let us single out the elements of the first column of each auxiliary matrix, $a_{ij \cdot j-1}$, $i \geqslant j$, designating them c_{ij}, $i \geqslant j$.

Analysing the process of computing the coefficients of the auxiliary matrices, we see that

$$
\begin{aligned}
a_{ij \cdot k} &= a_{ij \cdot k-1} - a_{ik \cdot k-1} b_{kj \cdot k-1} = a_{ij \cdot k-1} - c_{ik} b_{kj} \\
&= a_{ij \cdot k-2} - c_{ik-1} b_{k-1j} - c_{ik} b_{kj} = \cdots \\
&= a_{ij} - c_{i1} b_{1j} - c_{i2} b_{2j} - \cdots - c_{ik} b_{kj} = a_{ij} - \sum_{l=1}^{k} c_{il} b_{lj}.
\end{aligned}
$$

(1)

Thus any element $a_{ji \cdot k}$ is expressible in terms of an accumulation of the elements we have singled out, c_{ij}, and the numbers b_{ij}.

In particular, for the elements c_{ij}, $i \geqslant j$, and b_{ij}, $i < j$, themselves, the recurrence formulas

(2)

$$
c_{ij} = a_{ij \cdot j-1} = a_{ij} - \sum_{l=1}^{j-1} c_{il} b_{lj}; \quad i \geqslant j
$$

$$
b_{ij} = \frac{a_{ij \cdot i-1}}{a_{ii \cdot i-1}} = \frac{a_{ij} - \sum_{l=1}^{i-1} c_{il} b_{lj}}{c_{ii}}; \quad i < j
$$

hold. The constant terms of the transformed system are obviously to be determined in accordance with these same formulas.

The return course remains the same as in the full-length single-division scheme.

Computation by the compact scheme may be conveniently arranged as is shown in Table V.

Here we effect the computation of the elements c and b successively anglewise, beginning with the computation of the elements c:

c_{11}	b_{12}	b_{13}	b_{14}	1st step
c_{21}	c_{22}	b_{23}	b_{24}	2nd step
c_{31}	c_{32}	c_{33}	b_{34}	3rd step
c_{41}	c_{42}	c_{43}		

TABLE V.   *Compact Scheme for the Single-Division Method*

Symbolic scheme:

x_1	x_2	x_3	x_4		Σ
a_{11}	a_{12}	a_{13}	a_{14}	a_{15}	a_{16}
a_{21}	a_{22}	a_{23}	a_{24}	a_{25}	a_{26}
a_{31}	a_{32}	a_{33}	a_{34}	a_{35}	a_{36}
a_{41}	a_{42}	a_{43}	a_{44}	a_{45}	a_{46}
c_{11}	b_{12}	b_{13}	b_{14}	b_{15}	b_{16}
c_{21}	c_{22}	b_{23}	b_{24}	b_{25}	b_{26}
c_{31}	c_{32}	c_{33}	b_{34}	b_{35}	b_{36}
c_{41}	c_{42}	c_{43}	c_{44}	b_{45}	b_{46}
			1	x_4	$\bar{x}_4$
		1		x_3	$\bar{x}_3$
	1			x_2	$\bar{x}_2$
1				x_1	$\bar{x}_1$

Numerical values (staircase unit entries $=1$ in superior position on the diagonal):

x_1	x_2	x_3	x_4		Σ
1.00	0.42	0.54	0.66	0.3	2.92
0.42	1.00	0.32	0.44	0.5	2.68
0.54	0.32	1.00	0.22	0.7	2.78
0.66	0.44	0.22	1.00	0.9	3.22
1.00	0.42	0.54	0.66	0.3	2.92
0.42	0.82360 ^{1}	0.11316	0.19767	0.45410	1.76493 (5)
0.54	0.09320	0.69785 ^{1}	−0.22185*	0.71030*	1.48844 (40)
0.66	0.16280	−0.15482	0.49787 ^{1}	1.48240	2.48239 (40)
			1	1.48240	2.48239 (40)
		1		1.03917	2.03916 (7)
	1			0.04348	1.04348 (80)
1				−1.25780	−0.25779 (80)

Horizontal sums in superior position.

* Obtained by rounding to five decimals before performing division, which renders the procedure consistent with that of Table I. In practice this is not ordinarily bothered with, and is not here in the check column.

Here any element is obtained as the difference between the corresponding element a and the sum of paired products: of the elements c located in the given row (at the left) and the elements b located in the given column (above). Of course in computing the elements b it is still necessary to divide by the corresponding element c.

Thus

$$c_{43} = a_{43} - c_{41}b_{13} - c_{42}b_{23}$$

$$= 0.22 - 0.66 \cdot 0.54 - 0.16280 \cdot 0.11316 = -0.15482;$$

$$b_{34} = \frac{a_{34} - c_{31}b_{14} - c_{32}b_{24}}{c_{33}}$$

$$= \frac{0.22 - 0.54 \cdot 0.66 - 0.09320 \cdot 0.19767}{0.69785} = -0.22185.$$

We shall say a few words about the control to be applied when computing by the compact scheme. Just as before we form a column of check sums and perform the same operations upon it as upon the column of constant terms.

In this connection each number of the transformed column must coincide with the sum of the elements of the corresponding rows of the matrix B, augmented by adjoining the transformed column of constant terms. The matrix B is of course the matrix of the coefficients of the system obtained from the given one after completion of the forward course by the single-division scheme.[1]

In computing by the compact scheme it is convenient to use a triangle or two rulers in order to fix the attention on the required row and column.

---

[1] [Note that in Table V the units entered stepwise complete the tabulation of matrix B by filling in its diagonal; these of course figure in the row-total of the check. The return course, the third panel of Table V, is conducted by utilizing the same numbers $b_{ij.i-1}$ as were utilized in Tables I to III, in accordance with formulas (9), (5) and (3) of §6, i.e., analogous to formula (9b). The units arrayed northeast-southwest in the table serve, as there, the same purpose of locating the appropriate $b_{ij.i-1}$, these being found successively in successively higher rows of panel two here, as distinct from Tables I, II and III, where they were found at the bottom row of successively higher panels.]

§ 9. THE CONNECTION OF GAUSS'S METHOD WITH THE DECOMPOSITION OF A MATRIX INTO FACTORS

It was shown in § 1 that a system of n linear equations in n unknowns can be written in matrix form

$$(1) \qquad AX = F,$$

A being the given matrix, nonsingular, X and F being columns composed of the values of the unknowns and of the constant terms, respectively, and which we shall speak of as vectors.

The Gauss method, performed with a fixed order of leading elements, consists in replacing the given system by an equivalent triangular system by combining the equations linearly; this reduces to combining linearly the rows of A and F. In the course of its application, in using the single-division scheme, we are obliged to add to the elements of some rows elements proportional to the elements of the preceding rows, i.e., to effect upon matrix A elementary transformations of Type b' of Paragraph 10 of § 1.

The result of several transformations of this form, as was shown there, is equivalent to a premultiplication of the matrix A by some triangular matrix of the form

$$(2) \qquad \begin{pmatrix} \gamma_{11} & 0 & \ldots & 0 \\ \gamma_{21} & \gamma_{22} & \ldots & 0 \\ \cdot & \cdot & \cdot & \cdot \\ \gamma_{n1} & \gamma_{n2} & \ldots & \gamma_{nn} \end{pmatrix} = \Gamma.$$

As the result of these transformations we arrive at a system with a triangular matrix

$$(3) \qquad B = \begin{pmatrix} 1 & b_{12} & \ldots & b_{1n} \\ 0 & 1 & \ldots & b_{2n} \\ \cdot & \cdot & \cdot & \cdot \\ 0 & 0 & \ldots & 1 \end{pmatrix}.$$

Thus $\Gamma A = B$, which is to say $A = \Gamma^{-1} B$, and consequently the matrix A is factorable into the product of two triangular matrices.

The compact scheme realizes this factorization.   Indeed, $\Gamma^{-1}=C$, where the elements of the matrix C are defined by formulas (2) of § 8, for from these formulas it follows that

$$a_{ij} = c_{ij} + \sum_{l=1}^{j-1} c_{il}b_{lj} = \sum_{l=1}^{j} c_{il}b_{lj}, \quad i \geqslant j$$

(from the formulas for c_{ij}), and

$$a_{ij} = \sum_{l=1}^{i-1} c_{il}b_{lj} + c_{ii}b_{ij} = \sum_{l=1}^{i} c_{il}b_{lj}, \quad i < j$$

(from the formulas for b_{ij}).

These last formulas state that

$$A = CB.$$

Since the diagonal elements of the matrix B are equal to unity, such a factorization is unique.

The compact notation for the schemes of the Gauss method that differ from the single-division scheme are likewise related to the factorization of a matrix into the product of two triangular ones, but with another choice of the diagonal elements.

We remark that the compact schemes fix the order of the elimination, and they are therefore applicable only in case all the determinants a_{11}, $\begin{vmatrix} a_{11} & a_{12} \\ a_{21} & a_{22} \end{vmatrix}$, . . ., $|A|$ do not equal zero, as follows from the theorem of Paragraph 11, § 1.

We shall show that in case the matrix A is symmetric,

$$(4) \qquad\qquad b_{ik} = \frac{c_{ki}}{c_{ii}}.$$

We have, indeed, $A = CB$, $A' = B'C'$, and, since $A = A'$,

$$CB = B'C' = B' \begin{pmatrix} c_{11} & 0 & \cdots & 0 \\ 0 & c_{22} & \cdots & 0 \\ \cdot & \cdot & \cdot & \cdot \\ 0 & 0 & \cdots & c_{nn} \end{pmatrix} \begin{pmatrix} 1 & \frac{c_{21}}{c_{11}} & \cdots & \frac{c_{n1}}{c_{11}} \\ 0 & 1 & \cdots & \frac{c_{n2}}{c_{22}} \\ \cdot & \cdot & \cdot & \cdot \\ 0 & 0 & \cdots & 1 \end{pmatrix}.$$

Hence, on the strength of the uniqueness of the factorization of the matrix A into the product of the two triangular matrices, it follows that

$$
B = \begin{pmatrix}
1 & \dfrac{c_{21}}{c_{11}} & \cdots & \dfrac{c_{n1}}{c_{11}} \\
0 & 1 & \cdots & \dfrac{c_{n2}}{c_{22}} \\
& \cdot \; \cdot \; \cdot \; \cdot \; \cdot \; \cdot \\
0 & 0 & \cdots & 1
\end{pmatrix}.
$$

Thus the elements of B are simply the elements of the matrix C, divided. This notwithstanding, the compact scheme requires n^2 transcriptions even for the case of a symmetric A.

§ 10. THE SQUARE-ROOT METHOD [1]

In this section we shall show that in case the matrix of the system is symmetric, finding the solution may be rendered still easier, since in such case the matrix can be resolved into the product of two triangular matrices of which one is the transpose of the other.

Thus let

(1) $$A = S'S,$$

where

(2) $$
S = \begin{pmatrix}
s_{11} & s_{12} & \cdots & s_{1n} \\
0 & s_{22} & \cdots & s_{2n} \\
& \cdot \; \cdot \; \cdot \; \cdot \; \cdot \; \cdot \\
0 & 0 & \cdots & s_{nn}
\end{pmatrix}.
$$

Let us determine the elements of the matrix S. In view of the rule for the multiplication of matrices, we have

$$a_{ij} = s_{1i}s_{1j} + s_{2i}s_{2j} + \cdots + s_{ii}s_{ij}, \quad i < j$$

$$a_{ii} = s_{1i}^2 + s_{2i}^2 + \cdots + a_{ii}^2, \qquad i = j.$$

[1] T. Banachiewicz, [1].

Hence we obtain the formulas for the determination of the s_{ij}:

$$s_{11} = \sqrt{a_{11}}, \qquad s_{1j} = \frac{a_{1j}}{s_{11}},$$

$$(3) \quad s_{ii} = \sqrt{a_{ii} - \sum_{l=1}^{i-1} s_{li}^2}, \quad i > 1; \qquad s_{ij} = \frac{a_{ij} - \sum_{l=1}^{i-1} s_{li}s_{lj}}{s_{ii}}, \quad j > i;$$

$$s_{ij} = 0, \quad i > j.$$

Furthermore, the solution of the system reduces to the solution of two triangular systems. Indeed, the equation

$$AX = F,$$

is equivalent to the two equations

$$S'K = F, \qquad SX = K.$$

The elements of the vector K are determined by recurrence formulas analogous to the formulas for s_{ij}, to wit:

$$(4) \qquad k_1 = \frac{f_1}{s_{11}}, \quad k_i = \frac{f_i - \sum_{l=1}^{i-1} s_{li}k_l}{s_{ii}}; \qquad i > 1.$$

The final solution is found by the formulas

$$(5) \qquad x_n = \frac{k_n}{s_{nn}}, \quad x_i = \frac{k_i - \sum_{l=i+1}^{n} s_{il}x_l}{s_{ii}}; \qquad i < n.$$

In the scheme the usual control is to be employed, wherewith in forming the check elements we involve all the elements of the matrix. The control equation is an analog of that of the compact scheme:

$$\overline{k}_i = \sum_{k=1}^{n} s_{ik} + k_i.$$

With the square-root method, one has to record only the approximately $\frac{n^2}{2}$ elements of the matrix S and the $2n$ components of the vectors K and X.

We exhibit in Table VI the solution of our system by the square-root method.

TABLE VI.   The Square-Root Method

x_1	x_2	x_3	x_4	f	Σ						Σ
a_{11}	a_{12}	a_{13}	a_{14}	f_1	f_1	1.00	0.42	0.54	0.66	0.3	2.92
	a_{22}	a_{23}	a_{24}	f_2	f_2		1.00	0.32	0.44	0.5	2.68
		a_{33}	a_{34}	f_3	f_3			1.00	0.22	0.7	2.78
			a_{44}	f_4	f_4				1.00	0.9	3.22
s_{11}	s_{12}	s_{13}	s_{14}	k_1	k_1	1.00	0.42	$0.54\,^\bullet$	0.66	0.3	$2.92^{(2)}$
	s_{22}	s_{23}	s_{24}	k_2	k_2		0.90752	0.10270	0.17939	0.41211	1.60173
		s_{33}	s_{34}	k_3	k_3			0.83537*	−0.18533*	0.59336	1.24340
			s_{44}	k_4	k_4				0.70560	1.04597	1.75157
x_1	x_2	x_3	x_4			−1.25778	0.04348	1.03917	1.48238		
$\bar{x}_1$	$\bar{x}_2$	$\bar{x}_3$	$\bar{x}_4$			−0.25779	1.04349*	2.03917	2.48238		

Horizontal sums in superior position.

* Obtained by rounding radicand to five figures before extracting root, or by rounding numerator before division.

The computation of the elements s_{ij} (as also the elements $\bar{k}_i$ and k_i) is carried out by rows in sequence. Any diagonal element is calculated as the square root of the difference between the corresponding element a and the sum of the squares of all previously computed elements s located in the same column. A nondiagonal element s_{ij} is obtained by subtracting from the element a_{ij} the sum of the products, row by row, of corresponding elements s taken from the columns with numbers i and j. This difference, once obtained, is divided by the diagonal element of the row.

Thus

$$s_{34} = \frac{a_{34} - s_{13}s_{14} - s_{23}s_{24}}{s_{33}}$$

$$= \frac{0.22 - 0.54 \cdot 0.66 - 0.10270 \cdot 0.17939}{0.83537} = -0.18533.$$

The return course is carried through by the analogous formulas (5).

With a little experience, the computations, by formulas (3), (4) and (5), will go easily.

In case the elements of the matrix are such that radicands of the expression for s_{ii} are negative, one will encounter no substantial difficulties while using the square-root method. Indeed, in this case, in the row for which $s_{ii}^2 < 0$, only pure imaginary numbers figure; operations upon them are nowhit more complicated than upon real numbers.

Let us clarify this statement with an example

2	−1	1	4	6
	−2	3	5	5
		1	6	11
1.41421	−0.70711	0.70711	2.82843	4.24265
	1.58114i	−2.21360i	−4.42719i	−5.05965i
		2.32379	5.93858	8.26237
1.11112	0.77779	2.55556		
2.11112	1.77779	3.55556		

The square-root method is now widely employed where the solution of symmetric systems is called for; it can be recommended to the reader as one of the most efficient methods.

§ 11. THE INVERSION OF A MATRIX

As has already been remarked in the introduction, the problem of solving a nonhomogeneous linear system and that of inverting a matrix are closely connected with each other.

Indeed, if for the matrix A its inverse, A^{-1}, is known, then on multiplying the equation

$$(1) \qquad\qquad AX = F$$

on the left by A^{-1}, we obtain

$$(2) \qquad\qquad X = A^{-1}F.$$

Conversely, the determination of the elements of the inverse matrix may be reduced to the solution of n systems of the form

$$(3) \qquad \sum_{k=1}^{n} a_{ik}\alpha_{kj} = \delta_{ij} \qquad \begin{matrix} j = 1, \ldots, n \\[6pt] i = 1, \ldots, n, \end{matrix}$$

where δ_{ij} is Kronecker's symbol.

The last follows from the definition of the inverse matrix:

$$AA^{-1} = I$$

and the rule for matrix multiplication.

A technique employed in structural mechanics for determining the solution of a system by means of the so-called influence numbers[1] is none other than the solution of the system by construction of the inverse matrix. The influence numbers themselves are the elements of the inverse matrix.

The solution of the n systems necessary for the determination of the n^2 elements of the inverse matrix we shall perform by the scheme for several equations, as shown in Table VII. As the result we will obtain a matrix consisting of the rows of the inverse matrix, arranged

[1] A. A. Umansky [1].

in counter-order. For control of the computation and an estimate of the accuracy of the result, the multiplication of A by A^{-1} is expedient.

The theorem concerning the factorization of a matrix into the product of two triangular matrices makes possible the construction of a compact scheme for computing the elements of the inverse matrix.[1] The scheme requires $2n^2$ entries in all, of which n^2 entries give the elements of the inverse matrix.

Let

$$(4) \qquad A = CB,$$

where the elements of the triangular matrices C and B are determined by formulas (2) § 8, which we reproduce here:

$$c_{ij} = a_{ij} - \sum_{l=1}^{j-1} c_{il} b_{lj}, \quad i \geqslant j,$$

$$b_{ij} = \frac{a_{ij} - \sum_{l=1}^{i-1} c_{il} b_{lj}}{c_{ii}}, \quad i < j; \; b_{ii} = 1.$$

Let us denote the elements of the inverse matrix $A^{-1} = D$ by d_{ij}.

We have, obviously,

$$(5) \qquad D = B^{-1} C^{-1}.$$

We shall show that the elements d_{ij} can be determined without inverting the matrices B and C.

Multiplying equation (5) by C on the right, we obtain

$$(6) \qquad DC = B^{-1}.$$

The matrix B^{-1} is obviously also triangular, with units along the principal diagonal. We know, therefore, $\dfrac{n(n+1)}{2}$ of its elements: $\dfrac{n(n-1)}{2}$ will be null and the remaining n will be units.

Multiplying equation (5) by B on the left, we obtain, analogously,

$$(7) \qquad BD = C^{-1}.$$

---

[1] F. Waugh and P. Dwyer [1].

Since the matrix C^{-1} is triangular, $\dfrac{n(n-1)}{2}$ of its elements will be zeros.

It is readily seen that the system obtained by combining the $\dfrac{n(n+1)}{2}$ equations of the system $DC = B^{-1}$ that were mentioned above with the $\dfrac{n(n-1)}{2}$ equations of the system $BD = C^{-1}$ is a recurrent system that makes possible the determination of the n^2 elements of the inverse matrix.

We give it *in extenso* for $n = 4$:

$$
\begin{array}{lcccc}
 & i = 1 & 2 & 3 & 4 \\
c_{11}d_{i1} + c_{21}d_{i2} + c_{31}d_{i3} + c_{41}d_{i4} = & 1 & 0 & 0 & 0 \\
c_{22}d_{i2} + c_{32}d_{i3} + c_{42}d_{i4} = & & 1 & 0 & 0 \\
c_{33}d_{i3} + c_{43}d_{i4} = & & & 1 & 0 \\
c_{44}d_{i4} = & & & & 1
\end{array}
$$

$$
\begin{array}{lccc}
 & j = 2 & 3 & 4 \\
d_{1j} + b_{12}d_{2j} + b_{13}d_{3j} + b_{14}d_{4j} = & 0 & 0 & 0 \\
d_{2j} + b_{23}d_{3j} + b_{24}d_{4j} = & & 0 & 0 \\
d_{3j} + b_{34}d_{4j} = & & & 0 \,.
\end{array}
$$

From the equations of the first group for $i = 4$, d_{44}, d_{43}, d_{42}, d_{41} are determined in succession. Next d_{34}, d_{24}, d_{14} are determined from the equations of the second group for $j = 4$. Following this the process goes on analogously and we use the formulas of the first and second groups in turn. Explicitly, d_{31}, d_{32} and d_{33} are determined from the equations of the first group for $i = 3$, and d_{23} and d_{13} are determined from the equations of the second group for $j = 3$; d_{21} and d_{22} are determined from the equations of the first group for $i = 2$, and d_{12} from those of the second group for $j = 2$; finally, d_{11} is determined from the equations of the first group with $i = 1$. The inversion of the matrix by the compact scheme is exhibited in Table VIII.

Systems of Linear Equations

TABLE VII. *Inversion of a Matrix: Single-Division Scheme*

								Σ
1.00	0.42	0.54	0.66	1	0	0	0	3.62
0.42	1.00	0.32	0.44	0	1	0	0	3.18
0.54	0.32	1.00	0.22	0	0	1	0	3.08
0.66	0.44	0.22	1.00	0	0	0	1	3.32
1	0.42	0.54	0.66	1	0	0	0	3.62
	0.82360	0.09320	0.16280	-0.42	1	0	0	1.65960
	0.09320	0.70840	-0.13640	-0.54	0	1	0	1.12520
	0.16280	-0.13640	0.56440	-0.66	0	0	1	0.93080 [5]
	1	0.11316	0.19767	-0.50996	1.21418	0	0	2.01506
		0.69785	-0.15482	-0.49247	-0.11316	1	0	0.93740
		-0.15482	0.53222	-0.57698	-0.19767	0	1	0.60275 [6]
		1	-0.22185	-0.70570	-0.16216	1.43297	0	1.34327
			0.49787	-0.68624	-0.22278	0.22185	1	0.81072 [0]
			1	-1.37835	-0.44747	0.44560	2.00856	1.62838 [4]
		1		-1.01149	-0.26143	1.53183	0.44560	1.70453 [1]
	1			-0.12304	1.33221	-0.26142	-0.44746	1.50029 [3]
1				2.50759	-0.12303	-1.01149	-1.37834	0.99470

TABLE VIII. *Compact Scheme for Inverting a Matrix*

A

1.00	0.42	0.54	0.66
0.42	1.00	0.32	0.44
0.54	0.32	1.00	0.22
0.66	0.44	0.22	1.00

C, **B**, and $D = A^{-1}$

1.00 [1]	0.42	0.54	0.66	2.50759	−0.12304	−1.01149	−1.37834
0.42	0.82360 [1]	0.11316	0.19767	−0.12304	1.33221	−0.26142	−0.44746
0.54	0.09320	0.69785 [1]	−0.22185*	−1.01148	−0.26143	1.53183	0.44560
0.66	0.16280	−0.15482	0.49787 [1]	−1.37834	−0.44745	0.44560	2.00856

§ 12. THE PROBLEM OF ELIMINATION

This problem, in the simplest case, consists of the computation of the values of a linear form

$$(1) \qquad c_1 x_1 + c_2 x_2 + \cdots + c_n x_n,$$

where $c_1, c_2, \ldots, c_n$ are given numbers and $x_1, x_2, \ldots, x_n$ is the solution of the system

$$
\begin{aligned}
a_{11} x_1 + a_{12} x_2 + \cdots + a_{1n} x_n &= b_1 \\
a_{21} x_1 + a_{22} x_2 + \cdots + a_{2n} x_n &= b_2 \\
\cdots \cdots \cdots \cdots \cdots \cdots \cdots \\
a_{n1} x_1 + a_{n2} x_2 + \cdots + a_{nn} x_n &= b_n,
\end{aligned}
$$

(2)

whose determinant is different from zero.

The natural procedure for solving this problem consists in determining the numbers $x_1, x_2, \ldots, x_n$ in explicit form and substituting them in expression (1). This may be avoided, however, as follows.

Let us tabulate the matrix of the coefficients of system (2), adjoining the column of constant terms to the right of it, and underneath it a row consisting of the coefficients of the linear form to be computed, the signs of the coefficients being reversed. We shall put 0 in the lower right corner. We shall have the scheme

$$
(3) \qquad
\begin{array}{cccc||c}
a_{11} & a_{12} & \cdots & a_{1n} & b_1 \\
a_{21} & a_{22} & \cdots & a_{2n} & b_2 \\
\cdot & \cdot & \cdots & \cdot & \cdot \cdot \\
a_{n1} & a_{n2} & \cdots & a_{nn} & b_n \\
\hline
-c_1 & -c_2 & \cdots & -c_n & 0
\end{array}
$$

or, in abbreviated notation,

$$
(3') \qquad
\begin{array}{c||c}
A & b \\
\hline
-c & 0
\end{array}.
$$

Let $\gamma_1, \gamma_2, \ldots, \gamma_n$ designate the solution of the system of equations

$$a_{11}\gamma_1 + a_{21}\gamma_2 + \cdots + a_{n1}\gamma_n = c_1$$

$$a_{12}\gamma_1 + a_{22}\gamma_2 + \cdots + a_{n2}\gamma_n = c_2$$

(4)

$$\cdots \cdots \cdots \cdots \cdots$$

$$a_{1n}\gamma_1 + a_{2n}\gamma_2 + \cdots + a_{nn}\gamma_n = c_n.$$

Then

$$b_1\gamma_1 + b_2\gamma_2 + \cdots + b_n\gamma_n =$$

$$= (a_{11}x_1 + \cdots + a_{1n}x_n)\gamma_1$$

$$+ (a_{21}x_1 + \cdots + a_{2n}x_n)\gamma_2$$

$$+ \cdots \vdots \cdots \cdots$$

$$+ (a_{n1}x_1 + \cdots + a_{nn}x_n)\gamma_n$$

$$= (a_{11}\gamma_1 + \cdots + a_{n1}\gamma_n)x_1$$

$$+ (a_{12}\gamma_1 + \cdots + a_{n2}\gamma_n)x_2$$

$$+ \cdots \cdots \cdots \cdots$$

$$+ (a_{1n}\gamma_1 + \cdots + a_{nn}\gamma_n)x_n$$

$$= c_1x_1 + c_2x_2 + \cdots + c_nx_n.$$

Thus the computation of the form $c_1x_1 + c_2x_2 + \cdots + c_nx_n$ may be replaced by the computation of the form $b_1\gamma_1 + b_2\gamma_2 + \cdots + b_n\gamma_n$. On the other hand it is obvious that if we multiply the first row of scheme (3) by γ_1, the second by $\gamma_2, \ldots$, the nth by γ_n, and add to the last row, we will obtain a row whose elements situated beyond the double line are equal to zero; the element in the lower right corner obviously equals the number that is sought: the value of the linear form. Conversely, if we somehow choose a linear combination of n rows such that the addition of it to the last row gives a null row (up to the line), the coefficients of this combination will be solutions of system (4), and accordingly the element in the lower right corner will equal the sought number. This follows from the uniqueness of the solution of system (4). There is thus no need to find the numbers $\gamma_1, \gamma_2, \ldots, \gamma_n$. It is merely necessary to "annul the last row" by the addition of a suitable linear combination of the first n rows. This can be done by the ordinary forward course of the Gauss process, applied to scheme (3).

Just such a device may obviously be employed for the computation of the non-homogeneous linear form $c_1 x_1 + \cdots + c_n x_n + d$. There will be merely this difference, that in the lower right corner one must enter not 0 but the constant d, so that the initial scheme has the form

$$(3'') \qquad
\begin{array}{cccc||c}
a_{11} & a_{12} & \cdots & a_{1n} & b_1 \\
\cdot & \cdot & \cdots \cdot \cdot & \cdot & \cdot \cdot \\
a_{n1} & a_{n2} & \cdots & a_{nn} & b_n \\
\hline
-c_1 & -c_2 & \cdots & -c_n & d.
\end{array}$$

The solution of system (2) also may be found by this method, without employing a return course. Indeed, the expressions $x_1, x_2, \ldots, x_n$ are particular cases of linear form (1), with coefficients $(1, 0, \ldots, 0)$, $(0, 1, \ldots, 0), \ldots, (0, 0, \ldots, 1)$. Their determination may be effected simultaneously by the elimination scheme by simultaneously entering in the lower left corner the rows $(-1, 0, \ldots, 0)$, $(0, -1, \ldots, 0), \ldots, (0, 0, \ldots, -1)$, which collectively constitute the matrix $-I$. In the lower right corner, rather than the number 0, we place a null column.

The source scheme for the solution of the system thus has the form

$$(5) \qquad \begin{array}{c||c} A & b \\ \hline -I & 0 \end{array}.$$

Having obtained a null matrix in the lower left corner of the scheme by the addition of suitable linear combinations of the first n rows, we will have in the lower right corner a column of the values of the unknowns.

The inversion of a matrix is equivalent, as we have seen, to the solution of n systems of a special form, whose constant terms form the unit matrix. Their collective solution may be effected by means of the scheme

$$(6) \qquad \begin{array}{c||c} A & I \\ \hline -I & 0 \end{array},$$

where I, as before, designates the unit matrix, and in the lower right corner is arrayed a null matrix of the nth order. After annulling all the rows in the lower left corner, by the addition of suitable linear combinations of the first n rows, we will obtain the matrix A^{-1} in the lower right corner.

The result of the elimination may be cast in matrix form. This makes it possible to obtain some generalizations. To wit: the solution of the system, $x_1, \ldots, x_n$, may be written in matrix form as the column

$$
\begin{pmatrix} x_1 \\ x_2 \\ \vdots \\ x_n \end{pmatrix} = A^{-1}b,
$$

and the value of the linear form $c_1 x_1 + c_2 x_2 + \cdots + c_n x_n$ as the number[1] $cA^{-1}b$.

Such a representation points the way to the computation of the more complicated expression $CA^{-1}B$, where C and B are certain rectangular matrices, of which C consists of n columns and B of n rows, and the number of rows of C and columns of B is a matter of indifference. Indeed, the element of the ith row and kth column of the matrix $CA^{-1}B$ is the number $c_i A^{-1} b_k$, where c_i is the ith row of the matrix C, and b_k is the kth column of the matrix B.

The computation of the elements of the matrix $CA^{-1}B$ may therefore be carried out by the method of elimination, applied to the scheme

(7)
$$
\begin{array}{c|c} A & B \\ \hline -C & 0 \end{array}.
$$

After annulling the elements located in the lower left corner, by adding a linear combination of the first n rows, we obtain in the lower right corner the matrix $CA^{-1}B$.

[1] [It is to be remarked that the author's notation regards the row position as normal for a premultiplying vector, and the column position as normal for a postmultiplying one, no further diacritical marks thus being called for.]

Putting $C = I$, we obtain the scheme

(8)
$$\begin{array}{c|c} A & -B \\ \hline -I & 0 \end{array}$$

for the computation of the product $A^{-1}B$.

By transposing the scheme, one can obtain a modification of the method of elimination, to wit: if $D = CA^{-1}B$, then $D' = B'(A')^{-1}C'$; hence it follows that D' (and accordingly, D as well) may be computed by a forward course applied to the scheme

(9)
$$\begin{array}{c|c} A' & C' \\ \hline -B' & 0 \end{array}.$$

In particular, for computing the value of the linear form $c_1 x_1 + \cdots + c_n x_n$ one may also use the scheme

$$\begin{array}{c|c} A' & c' \\ \hline -b' & 0 \end{array},$$

where A' is the matrix that is the transpose of that of the coefficients of the system, c' is the column composed of the coefficients of the linear form being computed, b' is the row of constant terms, to be taken with signs reversed. The validity of this construction may be readily proved directly, without relying on the results set forth above, for if we multiply the first n rows of scheme (9) by $x_1, \ldots, x_n$ and add to the last, we will obtain zero in the lower left corner and $c_1 x_1 + \cdots + c_n x_n$ in the lower right corner.

Analogously, for solving the system $AX = b$ we may use the scheme

(10)
$$\begin{array}{c|c} A' & I \\ \hline -b' & 0 \end{array}.$$

We exhibit in Table IX the solution of the system, carried out by this modification of the method of elimination.

As usual, the last column of Table IX is the control; throughout the duration of the process each number obtained in it must coincide with the sum of the preceding numbers of the same row.

TABLE IX. *Solution of a System of Linear Equations by the Method of Elimination*

x_1	x_2	x_3	x_4					Σ
1.00	0.42	0.54	0.66	1	0	0	0	3.62
0.42	1.00	0.32	0.44	0	1	0	0	3.18
0.54	0.32	1.00	0.22	0	0	1	0	3.08
0.66	0.44	0.22	1.00	0	0	0	1	3.32
−0.3	−0.5	−0.7	−0.9	0	0	0	0	−2.40
1	0.42	0.54	0.66	1	0	0	0	3.62
	0.82360	0.09320	0.16280	−0.42	1	0	0	1.65960
	0.09320	0.70840	−0.13640	−0.54	0	1	0	1.12520
	0.16280	−0.13640	0.56440	−0.66	0	0	1	0.93080
	−0.37400	−0.53800	−0.70200	0.30	0	0	0	−1.31400 (5)
	1	0.11316	0.19767	−0.50996	1.21418	0	0	2.01506
		0.69785	−0.15482	−0.49247	−0.11316	1	0	0.93740
		−0.15482	0.53222	−0.57698	−0.19767	0	1	0.60275 (8)
		−0.49568	−0.62807	0.10927	0.45410	0	0	−0.56037 (6)
		1	−0.22185	−0.70570	−0.16216	1.43297	0	1.34327
			0.49787	−0.68624	−0.22278	0.22185	1	0.81072 (0)
			−0.73804	−0.24053	0.37372	0.71029	0	0.10546 (4)
			1	−1.37835	−0.44747	0.44560	2.00856	1.62838
				−1.25781	0.04347	1.03916	1.48240	1.30727

TABLE X. Computation of a Product $A^{-1}B$

1.00	0.42	0.54	0.66	1	0	0	0	3.62
0.42	1.00	0.32	0.44	0	1	0	0	3.18
0.54	0.32	1.00	0.22	0	0	1	0	3.08
0.66	0.44	0.22	1.00	0	0	0	1	3.32
−0.25	−0.45	−0.65	−0.85	0	0	0	0	−2.20
−0.30	−0.50	−0.70	−0.90	0	0	0	0	−2.40
−0.15	−0.30	−0.45	−0.60	0	0	0	0	−1.50
−0.20	−0.40	−0.60	−0.80	0	0	0	0	−2.00
1	0.42	0.54	0.66	1	0	0	0	3.62
	0.82360	0.09320	0.16280	−0.42	1	0	0	1.65960
	0.09320	0.70840	−0.13640	−0.54	0	1	0	1.12520
	0.16280	−0.13640	0.56440	−0.66	0	0	1	0.93080
	−0.34500	−0.51500	−0.68500	0.25	0	0	0	−1.29500
	−0.37400	−0.53800	−0.70200	0.30	0	0	0	−1.31400
	−0.23700	−0.36900	−0.50100	0.15	0	0	0	−0.95700
	−0.31600	−0.49200	−0.66800	0.20	0	0	0	−1.27600
	1	0.11316	0.19767	−0.50996	1.21418			2.01506 (5)

Matrix blocks as labeled in the table: A (upper-left 4×4), $-B'$ (below A), I (upper-right identity), O (zero block below I).

0.93740	0	1	−0.11316	−0.49247	−0.15482	0.69875
0.60275 (1)	1	0	−0.19767	−0.57698	0.53222	−0.15482
−0.59980 (8)	0	0	0.41889	0.07406	−0.61680	−0.47596
−0.56037	0	0	0.45410	0.10927	−0.62807	−0.49568
−0.47943 (5)	0	0	0.28776	0.02914	−0.45415	−0.34218
−0.63924	0	0	0.38368	0.03885	−0.60554	−0.45624
1.34327	0	1.43297	−0.16216	−0.70570	−0.22185	1
0.81072	1	0.22185	−0.22278	−0.68624	0.49787	
0.03954 (4)	0	0.68204	0.34171	−0.26182	−0.72239	
0.10546 (80)	0	0.71029	0.37372	−0.24053	−0.73804	
−0.01979 (40)	0	0.49033	0.23227	−0.21234	−0.53006	
−0.02639	0	0.65378	0.30970	−0.28312	−0.70676	
1.62838	2.00856	0.44560	−0.44747	−1.37835	1	
1.21587 (3)	1.45096	1.00394	0.01846	−1.25753		
1.30727 (2)	1.48240	1.03916	0.04347	−1.25781		
0.84335 (1)	1.06466	0.72652	−0.00492	−0.94295		
1.12448 (5)	1.41957	0.96871	−0.00655	−1.25728		

$(A^{-1}B)'$

Observation. In solving systems by the method of elimination the number of operations somewhat exceeds that for the Gauss method. The uniformity of the process, however, and especially the absence of a return course, not infrequently make this method the more convenient one.

For inverting a matrix the scheme

(11)
$$\begin{array}{c|c} A' & I \\ \hline -I & 0 \end{array}$$

may be used. Here, after completion of the process, we will obtain in the lower right corner a matrix that is the transpose of A^{-1}.

Thus for each of the problems analysed we have two schemes, defined either by the matrix A or by its transpose. We note that it is most efficient to use that scheme which contains the least number of rows in the lower left corner. Thus in the problem of solving a linear system without a return course it is most efficient to use the scheme with the transposed matrix; in the problem of computing the product $A^{-1}B$—the scheme with the transposed matrix in case the number of columns in B is less than n; the given matrix should be used as such if the number of columns in B is greater than n.

In Table X we exhibit the computation of the product $A^{-1}B$. Here, as A, we have taken the matrix

$$\begin{pmatrix} 1.00 & 0.42 & 0.54 & 0.66 \\ 0.42 & 1.00 & 0.32 & 0.44 \\ 0.54 & 0.32 & 1.00 & 0.22 \\ 0.66 & 0.44 & 0.22 & 1.00 \end{pmatrix},$$

and as B, the matrix

$$\begin{pmatrix} 0.25 & 0.30 & 0.15 & 0.20 \\ 0.45 & 0.50 & 0.30 & 0.40 \\ 0.65 & 0.70 & 0.45 & 0.60 \\ 0.85 & 0.90 & 0.60 & 0.80 \end{pmatrix}.$$

The computation of $A^{-1}B$ has been effected by scheme (9) with $C'=I$. We have as the result of the computations

$$A^{-1}B = \begin{pmatrix} -1.25753 & -1.25781 & -0.94295 & -1.25728 \\ 0.01846 & 0.04347 & -0.00492 & -0.00655 \\ 1.00394 & 1.03916 & 0.72652 & 0.96871 \\ 1.45096 & 1.48240 & 1.06466 & 1.41957 \end{pmatrix}.$$

§ 13. CORRECTION OF THE ELEMENTS OF THE INVERSE MATRIX

Inversion of a matrix by any of the schemes exhibited above offers no certainty as regards the accuracy of the results obtained, owing to the inevitable rounding errors, the influence of which on the final result is hard to gauge. For a check of the accuracy of the matrix D_0 obtained from the given matrix A by whichever process of inversion, one should determine the product AD_0; the discrepancy between it and the unit matrix will indicate the degree of inaccuracy in the results obtained.

Let this check computation reveal that the approximation D_0 to A^{-1} is such that $\|F_0\| \leqslant k < 1$, where

$$(1) \qquad\qquad F_0 = I - AD_0.$$

The first or second norm of those introduced in § 5 is conveniently taken as the norm of this matrix, they being the most readily computed.

Under this condition the elements of the inverse matrix, A^{-1}, may be computed to as high an accuracy as is desired by means of the following iterative process.[1]

---

[1] H. Hotelling, [1]. We note that Hotelling took as norm the expression $\left[\sum_{i=1, j=1}^{n} (a_{ij})^2 \right]^{\frac{1}{2}}$, for which the requirements of § 5 are not satisfied. In § 17 the relation between this expression and the third norm is explained.

Let us form the sequence of matrices

(2)
$$D_1 = D_0(I+F_0), F_1 = I-AD_1$$
$$D_2 = D_1(I+F_1), F_2 = I-AD_2$$
$$\cdot \quad \cdot \quad \cdot \quad \cdot \quad \cdot \quad \cdot \quad \cdot \quad \cdot \quad \cdot \quad \cdot \quad \cdot \quad \cdot$$
$$D_m = D_{m-1}(I+F_{m-1}), F_m = I-AD_m.$$

We shall show that the matrix $F_m = I-AD_m$ is equal to $F_0^{2^m}$. Indeed,

(3)
$$F_m = I-AD_m = I-AD_{m-1}(I+F_{m-1})$$
$$= I-(I-F_{m-1})(I+F_{m-1})$$
$$= F_{m-1}^2 = F_{m-2}^4 = \cdots = F_0^{2^m}.$$

Hence it follows that

(4)
$$D_m = A^{-1}(I-F_0^{2^m}).$$

The last formula shows that D_m approaches A^{-1}, the convergence of the process being very rapid. Let us make an estimate of the error, taking into consideration the fact that $A^{-1} = D_0(AD_0)^{-1} = D_0(I-F_0)^{-1}$:

(5)
$$\|D_m - A^{-1}\| = \|-A^{-1}F_0^{2^m}\| = \|-D_0(I-F_0)^{-1}F_0^{2^m}\|$$
$$\leqslant \|D_0\| \, \|(I-F_0)^{-1}\| \, \|F_0^{2^m}\| \leqslant \|D_0\| \frac{k^{2^m}}{1-k}.$$

It is seen from this estimate that, if only the initial approximation be so chosen that $\|I-AD_0\| \leqslant k < 1$, the number of correct figures of decimals increases in geometrical progression.

The successive approximations should be computed by lifting the parentheses in formulas (2), viz.:

(6)
$$D_m = D_{m-1}(I+F_{m-1}) = D_{m-1}+D_{m-1}(I-AD_{m-1}).$$

The second summand will here play the role of a small correction to the first.

Sometimes, starting with the matrix F_0, it is expedient to form the matrices $F_1 = F_0^2, F_2 = F_0^4 = (F_1)^2$, squaring the matrix F_0 successively, afterwards utilizing formula (6).

Observation. In case A is a symmetric matrix, and one has taken as the initial approximation D_0 a symmetric matrix, all the subsequent approximations will likewise be symmetric matrices, although the matrix F_0 may turn out to be non-symmetric. Indeed, from formula (6) it follows that $D_m = 2D_{m-1} - D_{m-1}AD_{m-1}$, whence, on assuming that $A' = A$, $D'_{m-1} = D_{m-1}$, we obtain

$$D'_m = 2D'_{m-1} - (D_{m-1}AD_{m-1})' = 2D'_{m-1} - D'_{m-1}A'D'_{m-1}$$
$$= 2D_{m-1} - D_{m-1}AD_{m-1} = D_m.$$

As an example, let us invert the matrix

$$\begin{pmatrix} 1.00 & 0.42 & 0.54 & 0.66 \\ 0.42 & 1.00 & 0.32 & 0.44 \\ 0.54 & 0.32 & 1.00 & 0.22 \\ 0.66 & 0.44 & 0.22 & 1.00 \end{pmatrix}.$$

As the initial approximation let us take from Table VII, § 11, the result of the inversion of the matrix A by the single-division method (retaining four figures after the decimal):

$$D_0 = \begin{pmatrix} 2.5076 & -0.1230 & -1.0115 & -1.3783 \\ -0.1230 & 1.3322 & -0.2614 & -0.4475 \\ -1.0115 & -0.2614 & 1.5318 & 0.4456 \\ -1.3783 & -0.4475 & 0.4456 & 2.0086 \end{pmatrix}.$$

The check computation gives, for F_0, the following value:

$$F_0 = 10^{-6} \cdot \begin{pmatrix} -52 & -18 & 20 & -50 \\ -60 & 8 & -10 & 10 \\ -18 & -34 & 26 & -10 \\ -66 & 20 & 10 & -54 \end{pmatrix}.$$

Hence we see that $\|F_0\|_I \leqslant 0.000150$, $\|F_0\|_{II} \leqslant 0.000196$. For the estimate of the error let us take $\|F_0\|_I$. On the strength of formula (5) we have, taking into account that $\|D_0\|_I < 5$:

$$\|D_1 - A^{-1}\|_I < 5 \, \frac{(0.00015)^2}{1 - 0.00015} < 0.0000001.$$

Thus D_1 gives for A^{-1} a value correct at least to a unit of the seventh decimal place for each of its elements.

Doing the computation, we obtain

$$D_1 = \begin{pmatrix} 2.50758616 & -0.12303930 & -1.01148870 & -1.37834207 \\ -0.12303930 & 1.33221281 & -0.26142705 & -0.44745375 \\ -1.01148870 & -0.26142705 & 1.53182667 & 0.44560858 \\ -1.37834207 & -0.44745375 & 0.44560858 & 2.00855152 \end{pmatrix}.$$

The check computation $I - AD_1$ gives

$$I - AD_1 = 10^{-8} \cdot \begin{pmatrix} 2 & 0 & 0 & 1 \\ 1 & 0 & -1 & 1 \\ 1 & 0 & 0 & 0 \\ 1 & 0 & 0 & -1 \end{pmatrix}.$$

§ 14. THE INVERSION OF A MATRIX BY PARTITIONING

It is sometimes expedient to partition a matrix before inversion. Let us examine the formulas for inverting a matrix of the nth order partitioned into four cells by the scheme

$$S = \left(\begin{array}{c|c} A & B \\ \hline C & D \end{array} \right),$$

where A and D are square matrices of orders p and q; $p + q = n$.

Let us seek the inverse matrix also in the form of a cellular matrix:

$$S^{-1} = \left(\begin{array}{c|c} K & L \\ \hline M & N \end{array} \right),$$

K and N being again square matrices of orders p and q.

In conformity with the rule for multiplying partitioned matrices, the following matrix equations must hold:

$$AK + BM = I$$
$$AL + BN = 0$$
$$CK + DM = 0$$
$$CL + DN = I.$$

Multiplying the third equation on the left by BD^{-1} and subtracting it from the first, we obtain

$$(A - BD^{-1}C)K = I,$$

whence

$$K = (A - BD^{-1}C)^{-1}.$$

Furthermore, from the third equation we find that

$$M = -D^{-1}CK.$$

In like manner we find from the second and fourth equations that

$$N = (D - CA^{-1}B)^{-1}$$

and

$$L = -A^{-1}BN.$$

Of course these formulas have been derived on the assumption that all the indicated matrix inversions are realizable.

The inversion of a matrix of order n thus reduces to the inversion of four matrices, of which two are of order p and two are of order q, and to several matrix multiplications.

The formulas developed above can be altered so that for the computation of the matrices K, L, M and N only two matrices, of orders p and q, need be inverted. For, as can readily be verified,

$$N = (D - CA^{-1}B)^{-1}, \quad M = -NCA^{-1},$$
$$L = -A^{-1}BN, \quad K = A^{-1} - A^{-1}BM,$$

and, analogously,

$$K = (A - BD^{-1}C)^{-1}, \quad L = -KBD^{-1},$$
$$M = -D^{-1}CK, \quad N = D^{-1} - D^{-1}CL.$$

The last formulas show that the method of partitioning is conveniently employed when any diagonal cell is easily inverted.

Let us take as an example the inversion of the matrix

$$\begin{pmatrix} 1.00 & 0.42 & : & 0.54 & 0.66 \\ 0.42 & 1.00 & : & 0.32 & 0.44 \\ \cdots & \cdots & : & \cdots & \cdots \\ 0.54 & 0.32 & : & 1.00 & 0.22 \\ 0.66 & 0.44 & : & 0.22 & 1.00 \end{pmatrix}.$$

The computation will be performed as follows:

1) We compute the matrix A^{-1}:

$$A^{-1} = \begin{pmatrix} 1.21418 & -0.50996 \\ -0.50996 & 1.21418 \end{pmatrix}$$

and form the products

$$A^{-1}B = \begin{pmatrix} 0.49247 & 0.57698 \\ 0.11316 & 0.19767 \end{pmatrix}, \quad CA^{-1} = \begin{pmatrix} 0.49247 & 0.11316 \\ 0.57698 & 0.19767 \end{pmatrix},$$

$$CA^{-1}B = \begin{pmatrix} 0.30215 & 0.37482 \\ 0.37482 & 0.46778 \end{pmatrix}.$$

By twice computing the last matrix, as $C(A^{-1}B)$ and $(CA^{-1})B$, we obtain a check on the preceding computations.

2) We form the matrix

$$(D-CA^{-1}B) = \begin{pmatrix} 0.69785 & -0.15482 \\ -0.15482 & 0.53222 \end{pmatrix}$$

and find its inverse

$$N = (D-CA^{-1}B)^{-1} = \begin{pmatrix} 1.53183 & 0.44560 \\ 0.44560 & 2.00855 \end{pmatrix}.$$

3) We compute the matrices

$$M = -NCA^{-1} = \begin{pmatrix} -1.01148 & -0.26142 \\ -1.37834 & -0.44745 \end{pmatrix}$$

$$L = -A^{-1}BN = \begin{pmatrix} -1.01148 & -1.37834 \\ -0.26142 & -0.44745 \end{pmatrix}$$

and

$$K = A^{-1} - A^{-1}BM = \begin{pmatrix} 2.50757 & -0.12305 \\ -0.12305 & 1.33221 \end{pmatrix}.$$

The sought inverse matrix will thus be:

$$S^{-1} = \begin{pmatrix} 2.50757 & -0.12305 & -1.01148 & -1.37834 \\ -0.12305 & 1.33221 & -0.26142 & -0.44745 \\ -1.01148 & -0.26142 & 1.53183 & 0.44560 \\ -1.37834 & -0.44745 & 0.44560 & 2.00855 \end{pmatrix}.$$

§ 15. THE BORDERING METHOD

In this section we shall consider computational schemes for the inversion of a matrix and the solution of a linear system that are based on the idea of bordering.

The given matrix A will be regarded as the result of bordering a matrix of the $(n-1)$th order, the inverse matrix of which will be considered to be known. Thus let

$$(1) \quad A_n = \begin{pmatrix} a_{11} & a_{12} & \cdots & a_{1,\,n-1} & \vdots & a_{1n} \\ a_{21} & a_{22} & \cdots & a_{2,\,n-1} & \vdots & a_{2n} \\ \cdot & \cdot & \cdots & \cdot & \vdots & \cdot \\ a_{n-1,\,1} & a_{n-1,\,2} & \cdots & a_{n-1,\,n-1} & \vdots & a_{n-1,\,n} \\ \cdots\cdots\cdots\cdots\cdots\cdots\cdots\cdots\cdots & \vdots & \cdots \\ a_{n1} & a_{n2} & \cdots & a_{n,\,n-1} & \vdots & a_{nn} \end{pmatrix}$$

$$= \begin{pmatrix} A_{n-1} & u_n \\ v_n & a_{nn} \end{pmatrix}.$$

Here A_{n-1} denotes the aforementioned matrix of the $(n-1)$th order,

$$(2) \qquad v_n = (a_{n1}, \ldots, a_{n,\,n-1}), \quad u_n = \begin{pmatrix} a_{1n} \\ \vdots \\ a_{n-1,\,n} \end{pmatrix}.$$

We seek the matrix A^{-1} as a bordered matrix too, in the form

$$(3) \qquad D_n = A_n^{-1} = \begin{pmatrix} P_{n-1} & r_n \\ q_n & \dfrac{1}{\alpha_n} \end{pmatrix},$$

where P_{n-1} is a matrix, q_n a row, r_n a column and $\dfrac{1}{\alpha_n}$ a number, all of which we have to determine.

By the rule for multiplication of bordered matrices we have

$$AA^{-1} = \begin{pmatrix} A_{n-1} & u_n \\ v_n & a_{nn} \end{pmatrix} \begin{pmatrix} P_{n-1} & r_n \\ q_n & \dfrac{1}{\alpha_n} \end{pmatrix}$$

$$= \begin{pmatrix} A_{n-1}P_{n-1} + u_n q_n, & A_{n-1}r_n + \dfrac{u_n}{\alpha_n} \\ v_n P_{n-1} + a_{nn}q_n, & v_n r_n + \dfrac{a_{nn}}{\alpha_n} \end{pmatrix} = \begin{pmatrix} I & 0 \\ 0 & 1 \end{pmatrix},$$

Hence

$$(4) \qquad A_{n-1}P_{n-1} + u_n q_n = I$$

$$(5) \qquad v_n P_{n-1} + a_{nn}q_n = 0$$

$$(6) \qquad A_{n-1}r_n + \frac{u_n}{\alpha_n} = 0$$

$$(7) \qquad v_n r_n + \frac{a_{nn}}{\alpha_n} = 1.$$

We have from equation (6)

$$(8) \qquad r_n = -\frac{A_{n-1}^{-1}u_n}{\alpha_n}.$$

Substituting this value of r_n in (7), we obtain

$$(9) \qquad \alpha_n = a_{nn} - v_n A_{n-1}^{-1} u_n.$$

Furthermore, we have from (4)

$$(10) \qquad P_{n-1} = A_{n-1}^{-1} - A_{n-1}^{-1}u_n q_n$$

and therefore, on the basis of (5) and (10):

$$v_n A_{n-1}^{-1} - v_n A_{n-1}^{-1} u_n q_n + a_{nn} q_n$$
$$= v_n A_{n-1}^{-1} - (a_{nn} - \alpha_n) q_n + a_{nn} q_n = v_n A_{n-1}^{-1} + \alpha_n q_n = 0.$$

Hence

(11)
$$q_n = -\frac{v_n A_{n-1}^{-1}}{\alpha_n}.$$

Finally,

(12).
$$P_{n-1} = A_{n-1}^{-1} + \frac{A_{n-1}^{-1} u_n v_n A_{n-1}^{-1}}{\alpha_n}.$$

Thus, at last,

(13)
$$A_n^{-1} = \begin{pmatrix} A_{n-1}^{-1} + \dfrac{A_{n-1}^{-1} u_n v_n A_{n-1}^{-1}}{\alpha_n}, & -\dfrac{A_{n-1}^{-1} u_n}{\alpha_n} \\ -\dfrac{v_n A_{n-1}^{-1}}{\alpha_n}, & \dfrac{1}{\alpha_n} \end{pmatrix},$$

where $\alpha_n = a_{nn} - v_n A_{n-1}^{-1} u_n$.

The constructed formula is obviously a particular case of the formulas for inverting a matrix by the partitioning method, p here equalling $n-1$ and q being 1.

The method developed here is basically a method of inverting a matrix by successive borderings: one constructs in succession the matrices inverse to the matrices

$$(a_{11}), \quad \begin{pmatrix} a_{11} & a_{12} \\ a_{21} & a_{22} \end{pmatrix}, \quad \begin{pmatrix} a_{11} & a_{12} & a_{13} \\ a_{21} & a_{22} & a_{23} \\ a_{31} & a_{32} & a_{33} \end{pmatrix}, \ldots$$

of which each successive one is obtained from the preceding by bordering. Each step of this process is accomplished on the basis of formula (13), viz.: if A_{n-1}^{-1} is already known, the following operations must be carried out in order to find A_n^{-1}:

1) The computation of the column: $-A_{n-1}^{-1} u_n$. The elements of the column, $\beta_{1n} \ldots \beta_{n-1, n}$, are found by accumulation.

2) The computation of the row: $-v_n A_{n-1}^{-1}$ with elements $\gamma_{n1} \cdots \gamma_{n, n-1}$.

3) The computation of the number

$$\alpha_n = a_{nn} + \sum_{i=1}^{n-1} a_{ni}\beta_{in} = a_{nn} + \sum_{i=1}^{n-1} a_{in}\gamma_{ni}.$$

(The computation twice of the number α_n is a good check on the preceding computations.)

4) Lastly, the determination of the elements d_{ik} of the inverse matrix by the formulas

(14)
$$d_{ik} = d'_{ik} + \frac{\gamma_{ni}\beta_{kn}}{\alpha_n} \quad (i, k \leqslant n-1)$$

$$d_{in} = \frac{\beta_{in}}{\alpha_n}, \qquad d_{nk} = \frac{\gamma_{nk}}{\alpha_n} \quad (i, k \leqslant n-1), \qquad d_{nn} = \frac{1}{\alpha_n}.$$

Here d'_{ik} are the elements of the matrix A_{n-1}^{-1}.

In case of a symmetric matrix the scheme is obviously abridged by exactly one half.

In Table XI the inversion of a matrix by the method of successive bordering is exhibited.

Each step of the process is set up on the following scheme:

(15)

$$\begin{array}{c|c||c}
A_{n-1}^{-1} & u_n & -A_{n-1}^{-1}u_n \\
\hline
v_n & a_{nn} & \\
-vA_{n-1}^{-1} & \vdots & \alpha_n
\end{array}$$

The method of bordering may also be useful in application to the solution of a system of linear equations. It is especially convenient to utilize this method in case it is necessary to solve a system for which the truncated system—obtained by striking out one equation and one unknown—has been solved earlier. Such a situation is often encountered in applications. For example, in solving problems of mathematical physics by the method of B. G. Galerkin or Ritz, it may occur that the solution obtained as the result of utilizing the $(n-1)$th coordinate function is insufficiently exact; if the addition of just one more coordinate function is sufficient for the construction of a more exact solution, the new system for the deter-

TABLE XI. *Inversion of a Matrix by Bordering*

Scheme:

1.00	0.42	0.54	0.66
0.42	1.00	0.32	0.44
0.54	0.32	1.00	0.22
0.66	0.44	0.22	1.00

A (center label)

Scheme box:

$$\begin{array}{c|c} A_{n-1}^{-1} & u_n \\ \hline v_n & a_{nn} \end{array} \quad -A_{n-1}^{-1}u_n = [\beta_{in}]$$

$$-v_nA_{n-1}^{-1} = [\gamma_{nj}] \qquad \alpha_n = a_{nn} - v_nA_{n-1}^{-1}u_n$$

1.00	0.42	−0.42
0.42	1.00	
−0.42		0.82360

Formulas:

$$\alpha_n = a_{nn} + \sum_{i=1}^{n-1} a_{ni}\beta_{in} = a_{nn} + \sum_{i=1}^{n-1} a_{in}\gamma_{ni}$$

$$A_{n-1}^{-1} = D_{n-1} = [d'_{ik}]$$

$$A_n^{-1} = D_n = [d_{ik}]$$

$$d_{ik} = d'_{ik} + \frac{\gamma_{ni}\beta_{kn}}{\alpha_n} \quad (i, k \leqslant n-1);$$

$$d_{in} = \frac{\beta_{in}}{\alpha_n}; \quad d_{nk} = \frac{\gamma_{nk}}{\alpha_n} \quad (i, k \leqslant n-1);$$

$$d_{nn} = \frac{1}{\alpha_n}$$

1.21418	−0.50996	0.54	−0.49247
−0.50996	1.21418	0.32	−0.11316
0.54	0.32	1.00	
−0.49247	−0.11316		0.69786

1.56171	−0.43010	−0.70569	0.66	−0.68623
−0.43010	1.23253	−0.16215	0.44	−0.22277
−0.70569	−0.16215	1.43295	0.22	0.22185
0.66	0.44	0.22	1.00	
−0.68623	−0.22277	0.22185		0.49788

2.50754	−0.12306	−1.01147	−1.37830
−0.12306	1.33221	−0.26141	−0.44744
−1.01147	−0.26141	1.53180	0.44559
−1.37830	−0.44744	0.44559	2.00852

A^{-1} (center label)

mination of the coefficients is obtained by bordering the previous system.

The method of bordering is applied to the solution of a system of linear equations in the following manner.

Let the system have the form

(16) $$A_n X_n = F_n.$$

Let us employ the symbols

(17) $$A_n = \begin{pmatrix} A_{n-1} & u_n \\ v_n & a_{nn} \end{pmatrix}; \quad F_n = \begin{pmatrix} F_{n-1} \\ f_n \end{pmatrix}; \quad X_n = \begin{pmatrix} y \\ x_n \end{pmatrix}.$$

Then

$$X_n = \begin{pmatrix} y \\ x_n \end{pmatrix} = \begin{pmatrix} A_{n-1}^{-1} & 0 \\ 0 & 0 \end{pmatrix} \begin{pmatrix} F_{n-1} \\ f_n \end{pmatrix}$$

$$+ \frac{1}{\alpha_n} \begin{pmatrix} A_{n-1}^{-1} u_n v_n A_{n-1}^{-1} & -A_{n-1}^{-1} u_n \\ -v_n A_{n-1}^{-1} & 1 \end{pmatrix} \begin{pmatrix} F_{n-1} \\ f_n \end{pmatrix}$$

$$= \begin{pmatrix} A_{n-1}^{-1} F_{n-1} \\ 0 \end{pmatrix} + \frac{1}{\alpha_n} \begin{pmatrix} A_{n-1}^{-1} u_n v_n A_{n-1}^{-1} F_{n-1} - A_{n-1}^{-1} u_n f_n \\ -v A_{n-1}^{-1} F_{n-1} + f_n \end{pmatrix}.$$

But $A_{n-1}^{-1} F_{n-1}$ is the solution of the truncated system, i.e., the system

$$a_{11} x_1 + \cdots + a_{1,\,n-1} x_{n-1} = f_1$$

$$\cdot \quad \cdot \quad \cdot \quad \cdot \quad \cdot \quad \cdot \quad \cdot \quad \cdot \quad \cdot \quad \cdot \quad \cdot \quad \cdot \quad \cdot$$

$$a_{n-1,\,1} x_1 + \cdots + a_{n-1,\,n-1} x_{n-1} = f_{n-1},$$

which solution we shall denote by X_{n-1}.

Analogously $A_{n-1}^{-1} u_n$ is the solution of that same system with the constant terms changed, that is, the solution of the system

$$a_{11} x_1 + \cdots + a_{1,\,n-1} x_{n-1} + a_{1,\,n} = 0$$

$$\cdot \quad \cdot \quad \cdot \quad \cdot \quad \cdot \quad \cdot \quad \cdot \quad \cdot \quad \cdot \quad \cdot \quad \cdot \quad \cdot \quad \cdot \quad \cdot$$

$$a_{n-1,\,1} x_1 + \cdots + a_{n-1,\,n-1} x_{n-1} + a_{n-1,\,n} = 0,$$

which solution we shall call Q_{n-1}.

Knowing X_{n-1} and Q_{n-1}, we easily compute X_n, for

(18)
$$X_n = \begin{pmatrix} X_{n-1} \\ 0 \end{pmatrix} + \frac{1}{a_{nn}+v_nQ_{n-1}} \begin{pmatrix} -Q_{n-1}v_nX_{n-1}+Q_{n-1}f_n \\ -v_nX_{n-1}+f_n \end{pmatrix}$$

$$= \begin{pmatrix} X_{n-1} \\ 0 \end{pmatrix} + \frac{f_n-v_nX_{n-1}}{a_{nn}+v_nQ_{n-1}} \begin{pmatrix} Q_{n-1} \\ 1 \end{pmatrix}.$$

Thus for the determination of X_n we must, in addition to X_{n-1}, still compute Q_{n-1}.

If the truncated system had been solved by Gauss's method, it would be necessary to have added yet another column in order to have found Q_{n-1}. The computation of the forward course is effected only for this column, utilizing the left half of the scheme, already known. Afterwards Q_{n-1} is determined by the usual return course.

§ 16. THE ESCALATOR METHOD

In inverting a matrix by the bordering method, the essential role in passing from the inversion of the matrix A_{k-1} to the inversion of the bordered matrix A_k is played by the computation of the expressions $-A_{k-1}^{-1}u_k$ and $-v_kA_{k-1}^{-1}$.

It is obvious that the components of the vector $-A_{k-1}^{-1}u_k$ are none other than the solution of the system of equations:

(0)
$$a_{11}z_{1k}+a_{12}z_{2k}+ \cdots +a_{1,\,k-1}z_{k-1,\,k}+a_{1k} = 0$$

$$\cdots \cdots \cdots \cdots \cdots \cdots \cdots \cdots$$

$$a_{k-1,\,1}z_{1k}+a_{k-1,\,2}z_{2k}+ \cdots +a_{k-1,\,k-1}z_{k-1,\,k}+a_{k-1,\,k} = 0.[1]$$

[1] $[-A_{k-1}^{-1}u_k$ of § 15 is a column vector with $(k-1)$ components, say z_k, where $z_k=\{z_{1k} \cdots z_{k-1,\,k}\}$ and thus

$$-A_{k-1}^{-1}u_k = z_k$$

$$u_k = -A_{k-1}z_k$$

i.e., $\qquad\qquad A_{k-1}z_k+u_k=0. \qquad (101)$

But $\qquad\qquad u_k = \{a_{1k} \cdots a_{k-1,\,k}\}.$

(101) is therefore the author's $(0)=(2)$ in matrix notation. Note that when $k=n+1$, (101) becomes the author's (1), with $z_{i,\,n+1}=x_i.]$

Analogously, the components of $-v_k A_{k-1}^{-1}$ form the solution of the transposed system.

The successive solution of such systems, for $k = 2, \ldots, n, n+1$, lies at the basis of the so-called escalator method for the solution of of systems of linear equations.[1] By the same token, the escalator method is closely connected with the method of bordering.

The author of the method limits himself to imparting a recipe for constructing the solution, and to formulas for the case $n = 4$.

The author considers the essential merit of the method to lie in the existence of a reliable check, by means of which the accuracy of the computations can be regulated; in particular, the method gives a result that is sufficiently reliable even in case the determinant of the coefficients of the system is small.

We shall set forth the escalator method only in application to systems with symmetric matrices, giving the computational scheme the compact form. The connection between the escalator method and Gauss's method will thereupon be revealed.

So let there be given a system of equations

$$(1) \quad \begin{aligned} a_{11}x_1 + a_{12}x_2 + \cdots + a_{1n}x_n + a_{1,\,n+1} &= 0 \\ \cdot \\ a_{n1}x_1 + a_{n2}x_2 + \cdots + a_{nn}x_n + a_{n,\,n+1} &= 0, \end{aligned}$$

the matrix of the coefficients of which is symmetric.

Designate by $z_{1k}, \ldots, z_{k-1,\,k}$ the solution of the system:

$$(2) \quad \begin{aligned} a_{11}z_{1k} + \cdots + a_{1,\,k-1}z_{k-1,\,k} + a_{1k} &= 0 \\ \cdot\ \cdot\ \cdot\ \cdot\ \cdot\ \cdot\ \cdot\ \cdot\ \cdot\ \cdot\ \cdot\ \cdot\ \cdot\ \cdot\ \cdot\ \cdot\ \cdot\ \cdot \\ a_{k-1,\,1}z_{1k} + \cdots + a_{k-1,\,k-1}z_{k-1,\,k} + a_{k-1,\,k} &= 0. \end{aligned}$$

The numbers $z_{i,\,n+1} = x_i$ obviously form the solution of system (1). If, therefore, we establish a means of computing in turn the numbers z_{ik} for $i < k \leqslant n+1$, we shall by that fact have given a method for constructing the sought solution.

Let us first assume that all the numbers z_{ik} have already been computed for $i < k \leqslant n$.

[1] J. Morris [1].

By means of them let us construct the matrix

$$(3) \qquad Z = \begin{pmatrix} 1 & z_{12} & z_{13} & \cdots & z_{1n} \\ 0 & 1 & z_{23} & \cdots & z_{2n} \\ 0 & 0 & 1 & \cdots & z_{3n} \\ \cdot & \cdot & \cdot & \cdot & \cdot \\ 0 & 0 & 0 & \cdots & 1 \end{pmatrix}.$$

It is easily seen that the matrix AZ has zeros above the principal diagonal. Indeed,

$$(4) \quad C_1 = AZ$$

$$= \begin{pmatrix} a_{11}, & a_{11}z_{12}+a_{12}, & \ldots, & a_{11}z_{1n}+a_{12}z_{2n}+\cdots+a_{1n} \\ a_{21}, & a_{21}z_{12}+a_{22}, & \ldots, & a_{21}z_{1n}+a_{22}z_{2n}+\cdots+a_{2n} \\ \cdot & \cdot & \cdot & \cdot \\ a_{n1}, & a_{n1}z_{12}+a_{n2}, & \ldots, & a_{n1}z_{1n}+a_{n2}z_{2n}+\cdots+a_{nn} \end{pmatrix}$$

and, in view of the definition of the numbers z_{ij}, all elements lying above the principal diagonal are equal to zero. The non-zero elements of the matrix C_1 are computed by the formulas

$$(5) \qquad c_{ij} = a_{i1}z_{1j}+a_{i2}z_{2j}+\cdots+a_{i,\,j-1}z_{j-1,\,j}+a_{ij}, \quad i \geqslant j.$$

Hence follows the connection between this method and that of Gauss, when the latter is considered in the light of a resolution of the matrix into factors.

Indeed, from (4) we have

$$(6) \qquad A = C_1 Z^{-1}.$$

But the matrix Z^{-1} is a triangular matrix with unit diagonal and zero elements below the diagonal; C_1 is a triangular matrix with zero elements above the diagonal. Comparing this resolution with the factorization of the matrix corresponding to the single-division scheme of the Gauss method ($A = CB$), we obtain, on the strength of the uniqueness of such a resolution, that

$$(7) \qquad Z^{-1} = B, \quad C_1 = C.$$

Let us represent the matrix C in the form

$$(8) \quad C = \begin{bmatrix} 1 & 0 & 0 & \ldots & 0 \\ \gamma_{21} & 1 & 0 & \ldots & 0 \\ \gamma_{31} & \gamma_{32} & 1 & \ldots & 0 \\ \cdot & \cdot & \cdot & \cdot & \cdot \\ \gamma_{n1} & \gamma_{n2} & \gamma_{n3} & \ldots & 1 \end{bmatrix} \begin{bmatrix} c_{11} & & & & \\ & c_{22} & & & \\ & & c_{33} & & \\ & & & \ddots & \\ & & & & c_{nn} \end{bmatrix} = \Gamma\Lambda.$$

Here

$$\gamma_{ij} = \frac{c_{ij}}{c_{jj}}, \quad i > j.$$

We shall show the matrix $\Gamma = C\Lambda^{-1}$ to be the matrix that is the transpose of the matrix Z^{-1}. In doing so we shall make essential use of the symmetry of the matrix A. We have :

$$A = \Gamma\Lambda Z^{-1} = A' = (Z^{-1})'\Lambda\Gamma'.$$

But $(Z^{-1})'$ is a triangular matrix with zeros above the diagonal; Γ' is a triangular matrix with zeros below the diagonal and units on it. On the strength of the uniqueness of the factorization, $\Gamma' = Z^{-1}$.

This circumstance permits us to give recurrence formulas for the successive determination of the numbers z_{ij}. Let us assume, indeed, that we have already computed the elements of the first k columns of the matrix Z. Then by formulas (5) the first k columns of the matrix C can be computed, and consequently the first k columns of the matrix Γ, that is, the first k rows of the matrix Z^{-1}. In order to continue the process we must compute the elements of the $(k+1)$th column of the matrix Z. Since the diagonal element of this column equals unity, and the elements below the diagonal equal zero, it remains for us to give formulas for the computation of $z_{i,\,k+1}$, where $i \leqslant k$. From the equation $Z\Gamma' = I$ we obtain, on the strength of the rule for the multiplication of matrices, the following recurrence formulas:

$$\gamma_{k+1,\,1} + \gamma_{k+1,\,2}z_{12} + \cdots + \gamma_{k+1,\,k}z_{1k} + z_{1,\,k+1} = 0$$
$$(9) \qquad \gamma_{k+1,\,2} + \cdots + \gamma_{k+1,\,k}z_{2k} + z_{2,\,k+1} = 0$$
$$\cdot \quad \cdot \quad \cdot \quad \cdot \quad \cdot \quad \cdot \quad \cdot \quad \cdot \quad \cdot \quad \cdot \quad \cdot$$
$$\gamma_{k+1,\,k} + z_{k,\,k+1} = 0,$$

defining the elements of the $(k+1)$th column of the matrix Z.

The reliable check alluded to earlier consists in the actual reduction to zero of the subdiagonal elements of the matrix C'.[1]

We subjoin the compact scheme of the escalator method for a system of four equations. (Recall that for uniformity of computation the constant terms are tabulated in the left parts of the equations.)

a_{11}	a_{12}	a_{13}	a_{14}	a_{15}	1	z_{12}	z_{13}	z_{14}	x_1
a_{21}	a_{22}	a_{23}	a_{24}	a_{25}	0	1	z_{23}	z_{24}	x_2
a_{31}	a_{32}	a_{33}	a_{34}	a_{35}	0	0	1	z_{34}	x_3
a_{41}	a_{42}	a_{43}	a_{44}	a_{45}	0	0	0	1	x_4
a_{15}	a_{25}	a_{35}	a_{45}						1

c_{11}	1	γ_{21}	γ_{31}	γ_{41}	γ_{51}
0		c_{22} 1	γ_{32}	γ_{42}	γ_{52}
0		0	c_{33} 1	γ_{43}	γ_{53}
0		0	0	c_{44} 1	γ_{54}
					1

[1] [Since $z_{n+1}=\{z_{1,\,n+1}\cdots z_{n+1,\,n+1}\}=\{x_1\ldots x_n\}=x$ is the sought vector, the computation is conducted with augmented matrices $\widehat{A}$ and $\widehat{Z}$ as follows. Denote system (2) by $Ax+h=0$ for brevity, and form the augmented matrices

$$\widehat{C}=\widehat{A}\widehat{Z}=\begin{bmatrix} A & \vdots & -h \\ \cdots & \vdots & \cdots \\ -h' & \vdots & 0 \end{bmatrix}\begin{bmatrix} Z & \vdots & x \\ \cdots & \vdots & \cdots \\ 0 & \vdots & 1 \end{bmatrix}=\begin{bmatrix} AZ & \vdots & Ax-h \\ \cdots & \vdots & \cdots \\ -h'Z & \vdots & -h'x \end{bmatrix}=\begin{bmatrix} C & \vdots & r \\ \cdots & \vdots & \cdots \\ -h'Z & \vdots & -h'x \end{bmatrix}$$

Here r is the vector of residuals, which must equal zero to the limit of computational accuracy; the trailing cell of C, $-h'x$, is not computed, being of no utility. Of $\widehat{C}$ it remains true as of C that all elements above the principal diagonal are zero, and therefore the reasoning that led to the equations for the determination of c_{ij}, γ_{ij} and z_{ij} remains valid. The following computor's formulas summarize the results:

Denoting by a_i the ith row of $\widehat{A}$,

$a_{(j}$ the jth column of $\widehat{A}$,

we have, since $\widehat{C}=\widehat{A}\widehat{Z}$

$$c_{ij}=a_i\,z_{(j}$$

(10)   So $c_{ij}=a_{(i}\,z_{(j}$, for $\widehat{A}$ is symmetric.

(11) $\gamma_{ij}=\dfrac{c_{ij}}{c_{jj}}$, $i>j$;

(12) z_{ij} is yielded by $z_i\,\gamma_{(j}=0$ for $i\neq j$.]

TABLE XII. *The Escalator Method: Compact Scheme*

$\hat{A}$ | | | | | $,\hat{Z}^{-1}$ | | | |
---|---|---|---|---|---|---|---|---|---
| 1.00 | 0.42 | 0.54 | 0.66 | -0.3 | 1 | -0.42 | -0.49247 | -0.68623 | -1.25776 |
| 0.42 | 1.00 | 0.32 | 0.44 | -0.5 | 0 | 1 | -0.11316 | -0.22277 | 0.04350 |
| 0.54 | 0.32 | 1.00 | 0.22 | -0.7 | 0 | 0 | 1 | 0.22185 | 1.03914 |
| 0.66 | 0.44 | 0.22 | 1.00 | -0.9 | 0 | 0 | 0 | 1 | 1.48237 |
| -0.3 | -0.5 | -0.7 | -0.9 | | 0 | 0 | 0 | 0 | 1 |

$\hat{C}'$ / Γ'				
1.00	0.42	0.54	0.66	-0.3
0	1.00 / 0.82360	0.11316	0.19767	-0.45410
0.000003	0.000003	1.00 / 0.69786	-0.22185	-0.71028
0.000006	-0.000005	-0.000001	1.00 / 0.49788	-1.48237
0.000010	-0.000009	-0.000009	-0.000001	1.00

Computing Formulas

1) $c_{ij} = a_i \, z_{ij}$ where a_i is the ith column of A.

2) $\gamma_{ij} = \dfrac{c_{ij}}{c_{jj}}$

3) z_{ij} is yielded by $z_i \gamma_j' = 0$ for $i \neq j$, where z_i is the ith row of Z

The scheme consists of three parts. In the first part is entered the matrix of the coefficients of the system (symmetric, under the condition). In the second part we gradually tabulate the elements of the columns of the matrix Z. In the lower part are entered the diagonal and subdiagonal elements of the matrix C' and the elements of the matrix Γ'. We note that the elements of the kth row of the matrix Γ' are obtained by multiplying the columns of the matrix A by the kth column of the matrix Z and dividing the sums obtained by the element c_{kk}. A column of matrix Z is filled up by formulas (9). In Table XII an illustrative example is given.

§ 17. THE METHOD OF ITERATION

We shall pass now to a description of iterative methods of solving systems of linear equations. These methods give the solution of the system in the form of the limit of a sequence of certain vectors, the construction of which is effected by a uniform process called the process of iteration.

Let the system of linear equations be given in the following form:

(1)
$$\begin{aligned}
x_1 &= a_{11}x_1 + a_{12}x_2 + \cdots + a_{1n}x_n + f_1 \\
x_2 &= a_{21}x_1 + a_{22}x_2 + \cdots + a_{2n}x_n + f_2 \\
&\cdots \cdots \cdots \cdots \cdots \cdots \cdots \\
x_n &= a_{n1}x_1 + a_{n2}x_2 + \cdots + a_{nn}x_n + f_n.
\end{aligned}$$

At first glance such a form of notation for the system may seem somewhat artificial, if only owing to the presence of like terms. This form is, however, especially convenient for the application of iterative methods. In the following section a number of devices will be considered which transform the given system into form (1). Every such device will give some modification of the iterative method.

Let us write system (1) in the form

(2) $$X = AX + F,$$

where A is the matrix of the coefficients, F the vector of constant members.

Let us construct the indicated sequence of vectors as follows: in the capacity of an initial approximation let us take a certain vector $X^{(0)}$, chosen, generally speaking, quite arbitrarily. Next let us construct the vectors

$$X^{(1)} = AX^{(0)} + F$$

$$X^{(2)} = AX^{(1)} + F$$

(3) $\cdot \quad \cdot \quad \cdot \quad \cdot \quad \cdot \quad \cdot \quad \cdot \quad \cdot \quad \cdot$

$$X^{(k)} = AX^{(k-1)} + F$$

$\cdot \quad \cdot \quad \cdot \quad \cdot \quad \cdot \quad \cdot \quad \cdot \quad \cdot$

If the sequence $X^{(0)}, X^{(1)}, \ldots, X^{(k)}, \ldots$ has the limit X, this limit will be the solution of system (2), for on passing to the limit in the equation $X^{(k)} = AX^{(k-1)} + F$ as $k \to \infty$, we obtain $X = AX + F$, which proves our statement.

We shall now give necessary and sufficient conditions of the convergence of the process of iteration, and also a number of sufficient conditions and estimates of the rate of convergence, based on the results of § 5.

THEOREM 1. *For convergence of the process of iteration with any initial vector $X^{(0)}$ and with any value of the vector F, it is necessary and sufficient that all proper numbers of the matrix A be less than unity in modulus.*

Proof.   One can readily convince oneself that

(4) $$X^{(k)} = A^k X^{(0)} + (I + A + \cdots + A^{k-1})F.$$

Indeed, for $k = 1$ this is true, and for the remaining k it is verifiable by the method of complete induction.   It hence follows that for the convergence of the iterative process with any F it is necessary that the series $I + A + \cdots + A^{k-1} + \cdots$ converge.   Indeed let us take as $X^{(0)}$ a null vector, and as F a vector all of whose components except the ith are equal to zero, the ith component being equal to unity. Then $X^{(k)}$ equals the ith column of the matrix $I + A + \cdots + A^{k-1}$. For convergence of the sequence $X^{(k)}$ it is necessary and sufficient that all elements of the ith column of the matrix $I + A + \cdots + A^{k-1}$ have a limit, and since i may be taken as anything from 1 to n, the series $I + A + \cdots + A^{k-1} + \cdots$ must converge.   But this condition is also sufficient, for with it satisfied, $A^k \to 0$, and consequently the first summand, $A^k X^{(0)}$, in equation (4) tends to zero, and the second summand, $(I+A+ \cdots +A^{k-1})F$ has a limit equal to $(I+A+ \cdots +A^{k-1}+ \cdots)F$. In § 5 it has been shown that the necessary and sufficient condition for convergence of the series $I+A+ \cdots +A^{k-1}+ \cdots$ is the inequality $|\lambda_i| < 1$, for all proper numbers λ_i of the matrix A. This proves Theorem 1.

Since the condition of Theorem 1 is difficult to verify, it is better to judge the convergence of the iterative process by means of some sufficient signs—those connected directly with the elements of the matrix. These criteria follow at once from Theorem 2, § 5.

THEOREM 2. *In order that the process of iteration converge, it is sufficient that any norm of the matrix A be less than unity.*

Proof. Indeed, if $\|A\| < 1$, all proper numbers of the matrix are less than unity and therefore, on the basis of Theorem 1, the iterative process converges.

We shall now give estimates of the rate of convergence of the iterative process in terms of the norm. Here the selection of the norm of vectors is a matter of complete indifference, but the matrix norm must be subordinate to the vector norm one has chosen.

THEOREM 3. *If $\|A\| < 1$, then*

$$(5) \qquad \|X - X^{(k)}\| \leqslant \|X^{(0)}\| \, \|A^k\| + \frac{\|F\| \, \|A\|^k}{1 - \|A\|}.$$

Proof. We have

$$\|X - X^{(k)}\|$$

$$= \|(I-A)^{-1}F - (I+A+ \cdots +A^{k-1})F - A^k X^{(0)}\|$$

$$\leqslant \|[(I-A)^{-1} - (I+A+ \cdots +A^{k-1})]F\| + \|A^k X^{(0)}\|$$

$$\leqslant \|(I-A)^{-1} - (I+A+ \cdots +A^{k-1})\| \, \|F\| + \|A\|^k \|X^{(0)}\|$$

$$\leqslant \|A\|^k \|X^{(0)}\| + \frac{\|F\| \, \|A\|^k}{1 - \|A\|}.$$

It is often important to compare the accuracy of two successive approximations, i.e., to compare the quantities $\|X - X^{(k)}\|$ and $\|X - X^{(k-1)}\|$.

Such a comparison may be made on the basis of the following theorem.

THEOREM 4. $\|X - X^{(k)}\| \leqslant \|A\| \, \|X - X^{(k-1)}\|.$

Proof.   Indeed, from the equations

$$X = AX + F$$

$$X^{(k)} = AX^{(k-1)} + F$$

it follows that

$$X - X^{(k)} = A(X - X^{(k-1)}).$$

Hence

$$\|X - X^{(k)}\| = \|A(X - X^{(k-1)}\| \leqslant \|A\| \, \|X - X^{(k-1)}\|.$$

The vector norms I, II, III, introduced by us in § 5, and the matrix norms subordinate to them, give the following easily tested sufficient tokens of convergence of the iterative process and estimates of the rate of its convergence.

I. If $\sum\limits_{j=1}^{n} |a_{ij}| \leqslant \mu < 1$, the process of iteration converges, and

$$(6) \qquad |x_i - x_i^{(k)}| \leqslant \mu \max_j |x_j - x_j^{(k-1)}|,$$

where $X = (x_1, \ldots, x_n)$ and $X^{(k)} = (x_1^{(k)}, \ldots, x_n^{(k)})$.

II. If $\sum\limits_{i=1}^{n} |a_{ij}| \leqslant \nu < 1$, the process of iteration converges, and

$$(7) \qquad \sum_{i=1}^{n} |x_i - x_i^{(k)}| \leqslant \nu \sum_{i=1}^{n} |x_i - x_i^{(k-1)}|.$$

III. If $\sum\limits_{\substack{i=1 \\ k=1}}^{n} a_{ik}^2 \leqslant P < 1$, the process of iteration converges, and

$$(8) \qquad \sqrt{\sum_{i=1}^{n} (x_i - x_i^{(k)})^2} \leqslant \sqrt{P} \sqrt{\sum_{i=1}^{n} (x_i - x_i^{(k-1)})^2}.$$

Indeed, the largest proper number λ_1 of the matrix $A'A$ does not exceed the sum of all the proper numbers of the matrix $A'A$, for they are all non-negative.

But

$$\lambda_1 + \lambda_2 + \cdots + \lambda_n = \operatorname{tr} A'A = \sum_{\substack{i=1 \\ k=1}}^{n} (a_{ik})^2 = \mathrm{P}.$$

Therefore $\|A\|_{\mathrm{III}} = \sqrt{\lambda_1} \leqslant \sqrt{P} < 1$, and consequently

$$\|X - X^{(k)}\|_{\mathrm{III}} = \sqrt{\sum_{i=1}^{n} (x_i - x_i^{(k)})^2} < \sqrt{P} \sqrt{\sum_{i=1}^{n} (x_i - x_i^{(k-1)})^2}.$$

We shall indicate still another way of constructing sufficient tokens of the convergence of the iterative process.

In the equation

$$(9) \qquad\qquad X = AX + F$$

with matrix

$$(10) \qquad\qquad A = \begin{pmatrix} a_{11} & a_{12} & \cdots & a_{1n} \\ \cdot & \cdot & \cdot & \cdot & \cdot & \cdot & \cdot \\ a_{n1} & a_{n1} & \cdots & a_{nn} \end{pmatrix}$$

let us introduce new unknowns $x_i = p_i z_i$, where p_i are some positive numbers.

System (1) will then be transformed into the system

$$(11) \qquad\qquad p_i z_i = \sum_{j=1}^{n} a_{ij} p_j z_j + f_i$$

or

$$(12) \qquad\qquad z_i = \sum_{j=1}^{n} a_{ij} \frac{p_j}{p_i} z_j + \frac{1}{p_i} f_i.$$

It is obvious that the components of the successive approximations $X^{(k)}$ for system (9) and $Z^{(k)}$ for system (12) will also be connected by the relations $x_i^{(k)} = p_i z_i^{(k)}$, provided these relations hold for the initial approximations $X^{(0)}$ and $Z^{(0)}$. The iterative processes for systems (9) and (12) therefore converge or diverge simultaneously, and consequently any sufficient condition for convergence of the iterative process for system (12) will also be a sufficient condition of convergence for system (9).

Thus if one can find positive numbers $p_1, \ldots, p_n$ such that one of the conditions

$$
\begin{aligned}
&1) \quad \sum_{j=1}^{n} |a_{ij}| \frac{p_j}{p_i} < 1 \quad \text{for } i = 1, \ldots, n \\
(13) \qquad &2) \quad \sum_{i=1}^{n} |a_{ij}| \frac{p_j}{p_i} < 1 \quad \text{for } j = 1, \ldots, n \\
&3) \quad \sum_{i,\,k=1}^{n} \frac{a_{ij}^2 p_j^2}{p_i^2} < 1,
\end{aligned}
$$

is satisfied, the process of iteration for system (9) will converge.

Observation. In the practical computation of the iterations we may proceed by two methods:

1) Put $X^{(0)} = 0$. Then

$$X^{(k)} = (I + A + \cdots + A^{k-1})F = F + AF + \cdots + A^{k-1}F.$$

For the computation of $X^{(k)}$ we calculate successively the vectors F, $AF, \ldots, A^{k-1}F$ and find their sum. This is convenient, because of the uniformity of the process of computation, and also because each succeeding term is only a correction to the sum of the preceding ones. The shortcoming of this method is the possible accumulation of rounding errors with the increase in the number of summands.

2) The computation is conducted directly by the formulas

$$X^{(k)} = AX^{(k-1)} + F.$$

Here each approximation is treated as a new initial one, and there is therefore no need in the first steps of the process to carry out the computations to any great accuracy; the errors that arise even themselves out later on.

As an example let us find the solution of the system

$$
(14) \quad
\begin{aligned}
0.78x_1 - 0.02x_2 - 0.12x_3 - 0.14x_4 &= 0.76 \\
-0.02x_1 + 0.86x_2 - 0.04x_3 + 0.06x_4 &= 0.08 \\
-0.12x_1 - 0.04x_2 + 0.72x_3 - 0.08x_4 &= 1.12 \\
-0.14x_1 + 0.06x_2 - 0.08x_3 + 0.74x_4 &= 0.68.
\end{aligned}
$$

Solving this system by the single-division scheme, we find

$$x_1 = 1.534965$$
$$x_2 = 0.122010$$
$$x_3 = 1.975156$$
$$x_4 = 1.412955.$$

For the application of the iterative process we bring the system into the form $X = AX + F$. It is possible to do this thus, for instance:

$$\text{(15)} \quad \begin{aligned} x_1 &= 0.22x_1 + 0.02x_2 + 0.12x_3 + 0.14x_4 + 0.76 \\ x_2 &= 0.02x_1 + 0.14x_2 + 0.04x_3 - 0.06x_4 + 0.08 \\ x_3 &= 0.12x_1 + 0.04x_2 + 0.28x_3 + 0.08x_4 + 1.12 \\ x_4 &= 0.14x_1 - 0.06x_2 + 0.08x_3 + 0.26x_4 + 0.68. \end{aligned}$$

It is readily apparent that the sufficient conditions for convergence of the process of iteration are fulfilled.

Let us compute the successive approximations by three methods, for a comparison of the course of the iterative process in the different variants, viz.:

1) The computation of the successive approximations is effected by the formula $X^{(k)} = \sum_{l=0}^{k} A^l F$ (see Table XIII).

2) The computation of the successive approximations is effected by the formula $X^{(k)} = AX^{(k-1)} + F$ with $X^{(0)} = F$ (see Table XIV).

3) Recompute $X^{(k)} = AX^{(k-1)} + F$ with $X^{(0)} = (1, 0, 0, 0)$ (see Table XV).

Table XIII will be explained. The first part of the table contains the components of the successively computed vectors $A^l F$. The last column is a check; in it are tabulated the numbers $\sum_{j=1}^{n} b_j x_j^{(k)}$, where $b_j = \sum_{i=1}^{n} a_{ij}$ (these numbers must be calculated beforehand and entered in the same sheet as that on which the elements of the matrix A are tabulated). But it is obvious that $\sum_{j=1}^{n} b_j x_j^{(k)} = \sum_{j=1}^{n} x_j^{(k+1)}$,

TABLE XIII. *Iterative Method: Computation*

by the Formula $X^{(k)} = \sum\limits_{l=0}^{k} A^l F$

1) $A^l F, l = 0, 1, \ldots, 14$

F	0.76	0.08	1.12	0.68	

	0.22	0.02	0.12	0.14	
	0.02	0.14 A'	0.04	0.06	
	0.12	0.04	0.28	0.08	
	0.14	-0.06	0.08	0.26	
	0.50	0.14 b	0.52	0.42	
F	0.76	0.08	1.12	0.68	
AF	0.3984	0.0304	0.4624	0.3680	1.2592
A^2F	0.1952 6400	0.0086 4000	0.2079 3600	0.1866 2400	0.5984 6400
A^3F	0.0942 1056	0.0022 3488	0.0969 2928	0.0919 7568	0.2853 5040
A^4F	0.0452 7913	0.0005 5572	0.0458 9292	0.0447 2340	0.1364 5117
A^5F	0.0217 4095	0.0001 3570	0.0218 8361	0.0216 0525	0.0653 6551 (5)
A^6F	0.0104 3649	0.0000 3285	0.0104 7017	0.0104 0364	0.0313 4316
A^7F	0.0050 0961	0.0000 0792	0.0050 1763	0.0050 0170	0.0150 3686 (79)
A^8F	0.0024 0463	0.0000 0190	0.0024 0654	0.0024 0272	0.0072 1580
A^9F	0.0011 5422	0.0000 0046	0.0011 5468	0.0011 5376	0.0034 6312 (20)
$A^{10}F$	0.0005 5403	0.0000 0011	0.0005 5414	0.0005 5392	0.0016 6219
$A^{11}F$	0.0002 6593	0.0000 0003	0.0002 6596	0.0002 6591	0.0007 9783
$A^{12}F$	0.0001 2765	0.0000 0001	0.0001 2765	0.0001 2764	0.0003 8295
$A^{13}F$	0.0000 6127		0.0000 6127	0.0000 6127	0.0001 8381
$A^{14}F$	0.0000 2941		0.0000 2941	0.0000 2941	0.0000 8823

2) $X^{(k)} = \sum\limits_{l=0}^{k} A^l F, \quad k = 12, 13, 14.$

$X^{(12)}$	1.5348 4720	0.1220 0958	1.9750 3858	1.4128 3762	
$X^{(13)}$	1.5349 0847	0.1220 0958	1.9750 9985	1.4128 9889	
$X^{(14)}$	1.5349 3788	0.1220 0958	1.9751 2926	1.4129 2830	

so that the elements of the check column are also equal to the sum of the rest of the elements lying in the same row. The second part of the table gives the approximate solution of the system, which we obtain by summing up the corresponding components of the computed vectors.

From a comparison of Tables XIII, XIV and XV with the result obtained by the single-division scheme, we see that the convergence

TABLE XIV. *Iterative Method: Computation by the Formula $X^{(k)} = AX^{(k-1)} + F$; $X^{(0)} = F$.*

	0.22	0.02	0.12	0.14	0.50
	0.02	0.14	0.04	−0.06	0.14
	0.12	0.04	0.28	0.08	0.52
	0.14	−0.06	0.08	0.26	0.42
	0.50	0.14	0.52	0.42	Σ
$F = X^{(0)}$	0.76	0.08	1.12	0.68	2.64
$X^{(1)}$	1.1584	0.1104	1.5824	1.0480	3.8992
$X^{(2)}$	1.3537	0.1190	1.7903	1.2346	4.4977 (6)
$X^{(3)}$	1.4479	0.1213	1.8873	1.3266	4.7830 (1)
$X^{(4)}$	1.4932	0.1218	1.9332	1.3713	4.9195
$X^{(5)}$	1.5149	0.1220	1.9551	1.3929	4.9849
$X^{(6)}$	1.5253	0.1220	1.9655	1.4033	5.0162 (1)
$X^{(7)}$	1.5303	0.1220	1.9705	1.4083	5.0312 (1)
$X^{(8)}$	1.53273	0.12201	1.97292	1.41072	5.03838
$X^{(9)}$	1.53389	0.12201	1.97408	1.41188	5.04187 (6)
$X^{(10)}$	1.53445	0.12201	1.97464	1.41244	5.04354
$X^{(11)}$	1.53472	0.12201	1.97491	1.41271	5.04434 (5)
$X^{(12)}$	1.53485	0.12201	1.97504	1.41284	5.04473 (4)
$X^{(13)}$	1.534910	0.122010	1.975101	1.412900	5.044920 (1)
$X^{(14)}$	1.5349385	0.1220096	1.9751299	1.4129289	5.0450069

of the process in all three variants is nearly identical: the fourteenth
approximation gives, in the example in hand, a valid result with
accuracy to a unit of the fourth decimal place.

The convergence of the iterative process can be much improved
by means of a simple device suggested by L. A. Liusternik [1].

TABLE XV.   *Iterative Method: Computation by*
the Formula $X^{(k)} = AX^{(k-1)} + F$; $X^{(0)} = (1, 0, 0, 0)$

	0.22	0.02	0.12	0.14	0.50
	0.02	0.14	0.04	−0.06	0.14
	0.12	0.04	0.28	0.08	0.52
	0.14	−0.06	0.08	0.26	0.42
	0.50	0.14	0.52	0.42	Σ
F	0.76	0.08	1.12	0.68	2.64
$X^{(0)}$	1	0	0	0	
$X^{(1)}$	0.98	0.10	1.24	0.82	3.14
$X^{(2)}$	1.2412	0.1140	1.6544	1.1236	4.1332
$X^{(3)}$	1.3912	0.1195	1.8266	1.2714	4.6088 (7)
$X^{(4)}$	1.4656	0.1213	1.9049	1.3443	4.8362 (1)
$X^{(5)}$	1.5016	0.1218	1.9416	1.3798	4.9449 (8)
$X^{(6)}$	1.5190	0.1220	1.9591	1.3970	4.9970 (1)
$X^{(7)}$	1.5273	0.1220	1.9675	1.4053	5.0221
$X^{(8)}$	1.53129	0.12201	1.97148	1.40928	5.03406
$X^{(9)}$	1.53320	0.12201	1.97339	1.41119	5.03979
$X^{(10)}$	1.53412	0.12201	1.97431	1.41211	5.04254 (5)
$X^{(11)}$	1.53456	0.12201	1.97475	1.41255	5.04387
$X^{(12)}$	1.534770	0.122010	1.974962	1.412761	5.044502 (3)
$X^{(13)}$	1.534872	0.122010	1.975063	1.412862	5.044806 (7)
$X^{(14)}$	1.534920	0.122010	1.975112	1.412911	5.044952 (3)

In view of the fact that this device is connected with the approximate computation of the first proper number of the matrix, we postpone consideration of it until § 35, in Chapter III.

§ 18. THE PREPARATORY CONVERSION OF A SYSTEM OF LINEAR EQUATIONS INTO FORM SUITABLE FOR THE METHOD OF ITERATION

In the preceding section we have established a number of sufficient conditions for the convergence of the process of iteration, as applied to systems of equations given in the form

$$(1) \qquad\qquad X = AX + F$$

or, what is the same thing,

$$(I - A)X = F.$$

All those criteria required that the matrix A be small in one sense or another, i.e., that in the matrix $(I-A)$ the diagonal elements be sufficiently preponderant and be close to unity.

Let us consider some devices for transforming a given system $BX = G$ into form (1) with a suitable matrix A.

A system

$$(2) \qquad\qquad BX = G$$

is especially easily transformed into the necessary form in case the diagonal elements of the matrix B considerably preponderate over the remaining elements.

Let us rewrite system (2) in extended form:

$$(3) \qquad \begin{aligned} b_{11}x_1 + b_{12}x_2 + \cdots + b_{1n}x_n &= g_1 \\ b_{21}x_1 + b_{22}x_2 + \cdots + b_{2n}x_n &= g_2 \\ \cdot \quad \cdot \quad \cdot \quad \cdot \quad \cdot \quad \cdot \quad \cdot \quad \cdot \quad \cdot \quad \cdot \\ b_{n1}x_1 + b_{n2}x_2 + \cdots + b_{nn}x_n &= g_n. \end{aligned}$$

Divide each equation of system (3) by the diagonal element. We obtain the system

$$x_1 + \frac{b_{12}}{b_{11}} x_2 + \cdots + \frac{b_{1n}}{b_{11}} x_n = \frac{g_1}{b_{11}}$$

(4)
$$\frac{b_{21}}{b_{22}} x_1 + x_2 + \cdots + \frac{b_{2n}}{b_{22}} x_n = \frac{g_2}{b_{22}}$$

$$\cdots \cdots \cdots \cdots \cdots \cdots \cdots$$

$$\frac{b_{n1}}{b_{nn}} x_1 + \frac{b_{n2}}{b_{nn}} x_2 + \cdots + x_n = \frac{g_n}{b_{nn}}.$$

This system, in matrix form, is given notationally by

(5) $$X = AX + F,$$

where

(6) $$A = \begin{pmatrix} 0 & -\dfrac{b_{12}}{b_{11}} & \cdots & -\dfrac{b_{1n}}{b_{11}} \\ -\dfrac{b_{21}}{b_{22}} & 0 & \cdots & -\dfrac{b_{2n}}{b_{22}} \\ \cdots & \cdots & \cdots & \cdots \\ -\dfrac{b_{n1}}{b_{nn}} & -\dfrac{b_{n2}}{b_{nn}} & \cdots & 0 \end{pmatrix}, \quad F = \begin{pmatrix} \dfrac{g_1}{b_{11}} \\ \dfrac{g_2}{b_{22}} \\ \vdots \\ \dfrac{g_n}{b_{nn}} \end{pmatrix}.$$

For application of the process of iteration there is no need to actually make the transformation of (3) into system (5). The successive approximations may be computed by the formulas

(7)
$$b_{11} x_1^{(k)} = g_1 - b_{12} x_2^{(k-1)} - \cdots - b_{1n} x_n^{(k-1)}$$

$$\cdots \cdots \cdots \cdots \cdots \cdots \cdots \cdots$$

$$b_{nn} x_n^{(k)} = g_n - b_{n1} x_1^{(k-1)} - \cdots - b_{nn-1} x_{n-1}^{(k-1)}.$$

The modification of the iterative process just described bears the appellation *method of simple iteration*.

The sufficient conditions of convergence of the iterative process that were deduced in § 17, on being applied to system (5), give the following sufficient conditions for the method of simple iteration:

$$\text{I.} \quad \sum_{j=1}^{n}{}' \left| \frac{b_{ij}}{b_{ii}} \right| < 1$$

(8) $$\text{II.} \quad \sum_{i=1}^{n}{}' \left| \frac{b_{ij}}{b_{ii}} \right| < 1$$

$$\text{III.} \quad \sum_{i,j=1}^{n}{}' \left(\frac{b_{ij}}{b_{ii}} \right)^2 < 1.$$

Here the prime sign on the sum signifies that the value $i=j$ is omitted in the summation.

If one makes use of the generalized criteria (13) § 17, having put $p_i = \dfrac{1}{|b_{ii}|}$, we shall then obtain the following additional sufficient tokens of the convergence of the simple iteration:

$$\text{I}'. \quad \sum_{j=1}^{n}{}' \left| \frac{b_{ij}}{b_{jj}} \right| < 1$$

(8') $$\text{II}'. \quad \sum_{i=1}^{n}{}' \left| \frac{b_{ij}}{b_{jj}} \right| < 1$$

$$\text{III}'. \quad \sum_{i,j=1}^{n}{}' \left(\frac{b_{ij}}{b_{jj}} \right)^2 < 1.$$

The transformation of system (2) into system (5) just mentioned is obviously equivalent to the multiplication of system (2) on the left by the matrix

(9) $$D = \begin{pmatrix} \frac{1}{b_{11}} & & & \\ & \frac{1}{b_{22}} & & 0 \\ & & \ddots & \\ & 0 & & \frac{1}{b_{nn}} \end{pmatrix}.$$

We thereby pass to the system $DBX=DG$, which we write in the form $X=(I-DB)X+DG$.

Assume now that a system

$$BX = G$$

has been given, in which a predominance of the principal diagonal does not obtain. For application of the iterative process this system must be transformed into an equivalent system of the form

$$(I-A)X = F,$$

with the matrix $I-A$ having a predominant principal diagonal. For this it is enough to premultiply both sides of the equation $BX=G$ by an auxiliary matrix D sufficiently close to the matrix B^{-1}. Indeed, after such a multiplication we will obtain the system

$$DBX = DG,$$

in which the matrix DB is close to $B^{-1}B=I$. Let $DB=I-A$, $DG=F$. Since the matrix A will be small here, this transformation will bring the given system into the required form.

The selection of the auxiliary matrix D may be accomplished by various means, for instance by a rough inversion of the matrix B by the Gauss method. In case the predominance of the diagonal elements in the original system $BX=G$ does exist, but is insufficient, it is usually expedient to take as D the matrix inverse to the matrix

$$\begin{pmatrix} b_{11} & b_{12} & & & & \\ b_{21} & b_{22} & & & 0 & \\ & & b_{33} & b_{34} & & \\ & & b_{43} & b_{44} & & \\ & & & & \cdot & \\ 0 & & & & & \cdot \end{pmatrix}.$$

The inversion of such a matrix presents no difficulty, for it reduces to the inversion of matrices of the second order, viz.:

$$D = \begin{pmatrix} \dfrac{b_{22}}{\Delta_1} & -\dfrac{b_{12}}{\Delta_1} & \\ -\dfrac{b_{21}}{\Delta_1} & \dfrac{b_{11}}{\Delta_1} & \\ & & \ddots \end{pmatrix},$$

where Δ_1 is the determinant $\begin{vmatrix} b_{11} & b_{12} \\ b_{21} & b_{22} \end{vmatrix}$.

We note that the necessary and sufficient condition for the convergence of the iterations in application to the prepared system

$$DBX = DG$$

consists in all proper numbers of the matrix $I - DB$ being less than unity.

§ 19. SEIDEL'S METHOD

Let a system of linear equations be given in the form

(1) $$X = AX + F,$$

where A is the given matrix, F the given vector and X the sought vector with components $(x_1, \ldots, x_n)$.

The iterative method of Seidel is reminiscent of the ordinary iterative process, with this difference, that in computing the kth approximation to the component x_i, one takes into consideration the kth approximations, already computed, to the components $x_1, \ldots, x_{i-1}$. Explicitly, the computation of the successive approximations is performed by the formulas

(2) $$x_i^{(k)} = \sum_{j=1}^{i-1} a_{ij} x_j^{(k)} + \sum_{j=i}^{n} a_{ij} x_j^{(k-1)} + f_i$$

(instead of by $x_i^{(k)} = \sum_{j=1}^{n} a_{ij} x_j^{(k-1)}$ with the method of iteration).

In a number of cases it turns out that the Seidel process converges faster than the ordinary process of iteration.

Such is the case, in particular, on condition that

$$\|A\|_1 = \max_i \sum_{j=1}^n |a_{ij}| \leqslant \mu < 1.$$

Indeed, as we have seen, in this case there holds for the method of iteration the estimate

$$(3) \qquad \|X - X^{(k)}\| \leqslant \mu \|X - X^{(k-1)}\|,$$

where as the norm of vectors the first norm is taken, i.e., $\max_i |x_i|$.

We shall show that under this condition the Seidel process converges, and that a somewhat better estimate obtains for it.

Indeed, if $x_1, \ldots, x_n$ is the solution of system (1), then

$$(4) \qquad x_i = \sum_{j=1}^n a_{ij} x_j + f_i, \quad i = 1, \ldots n.$$

Subtracting equation (2) from (4), we obtain

$$x_i - x_i^{(k)} = \sum_{j=1}^{i-1} a_{ij}(x_j - x_j^{(k)}) + \sum_{j=i}^n a_{ij}(x_j - x_j^{(k-1)}),$$

whence it follows that

$$(5) \qquad |x_i - x_i^{(k)}| \leqslant \sum_{j=1}^{i-1} |a_{ij}| \, |x_j - x_j^{(k)}| + \sum_{j=i}^n |a_{ij}| \, |x_j - x_j^{(k-1)}|.$$

Let us employ the symbols

$$\sum_{j=1}^{i-1} |a_{ij}| = \beta_i, \quad \sum_{j=i}^n |a_{ij}| = \gamma_i.$$

Then from inequality (5) it follows that

$$|x_i - x_i^{(k)}| \leqslant \beta_i \|X - X^{(k)}\| + \gamma_i \|X - X^{(k-1)}\|.$$

Taking for i that value i_0 for which $|x_i - x_i^{(k)}|$ attains its maximum, we obtain

$$(6) \qquad \|X - X^{(k)}\| \leqslant \frac{\gamma_{i_0}}{1 - \beta_{i_0}} \|X - X^{(k-1)}\|,$$

for $\|X - X^{(k)}\| = \max_i |x_i - x_i^{(k)}| = |x_{i_0} - x_{i_0}^{(k)}|.$

Let us use the designation

(7) $$\max \frac{\gamma_i}{1 - \beta_i} = \mu'.$$

Then

(8) $$\|X - X^{(k)}\| \leqslant \mu' \|X - X^{(k-1)}\|.$$

We shall establish that

(9) $$\mu' \leqslant \mu.$$

Indeed,

$$\sum_{j=1}^{n} |a_{ij}| = \beta_i + \gamma_i$$

and

$$\beta_i + \gamma_i - \frac{\gamma_i}{1 - \beta_i} = \frac{\beta_i(1 - \beta_i - \gamma_i)}{1 - \beta_i} \geqslant 0,$$

for $\beta_i + \gamma_i \leqslant \mu < 1$. Hence

$$\mu = \max (\beta_i + \gamma_i) \geqslant \max \frac{\gamma_i}{1 - \beta_i} = \mu'.$$

Here the equality sign is possible only if $\max\limits_{i} \sum\limits_{j=1}^{n} |a_{ij}|$ is attained for $i = 1$, and the reduction of the estimate in comparison with estimate (3) will be best if the equations be arranged in increasing order of the $\sum\limits_{j=1}^{n} |a_{ij}|$, adopting as the first that equation in which this sum is least.

The Seidel method nevertheless does not always prove to have the advantage over the ordinary iterative process.

Sometimes the Seidel process converges more slowly than the process of iteration.   It is even possible that the Seidel method may diverge while the method of iteration converges.   The regions of convergence of these two processes are different, overlapping only partially.

We shall now establish the relation between the Seidel method and the method of iteration, which will make it possible to indicate the necessary and sufficient condition of convergence of the Seidel process.

Let us represent the equation

$$X = AX + F$$

in the form

(10) $$X = (B+C)X + F,$$

where

(11)

$$B = \begin{pmatrix} 0 & 0 & \cdots & 0 & 0 \\ a_{21} & 0 & \cdots & 0 & 0 \\ \cdot & \cdot & \cdot & \cdot & \cdot \\ a_{n1} & a_{n2} & \cdots & a_{n,\,n-1} & 0 \end{pmatrix}, \quad C = \begin{pmatrix} a_{11} & a_{12} & \cdots & a_{1n} \\ 0 & a_{22} & \cdots & a_{2n} \\ \cdot & \cdot & \cdot & \cdot \\ 0 & 0 & \cdots & a_{nn} \end{pmatrix}.$$

With these symbols one may represent the formulas

$$x_i^{(k)} = \sum_{j=1}^{i-1} a_{ij} x_j^{(k)} + \sum_{j=i}^{n} a_{ij} x_j^{(k-1)} + f_i$$

in matrix form as

(12) $$X^{(k)} = BX^{(k)} + CX^{(k-1)} + F.$$

Hence it follows that

$$X^{(k)} = (I-B)^{-1} C X^{(k-1)} + (I-B)^{-1} F.$$

Thus the Seidel method for system (10) is equivalent to the application of the method of iteration to the system

(13) $$X = (I-B)^{-1} CX + (I-B)^{-1} F,$$

which is equivalent to an initial system

$$X = (B+C)X + F$$

and may be obtained from it by premultiplication by the non-singular matrix $(I-B)^{-1}$.

From such a representation of the Seidel method it follows that for its convergence it is necessary and sufficient that all proper numbers of the matrix $M = (I-B)^{-1}C$ be less than unity in modulus. These proper numbers are the roots of the equation $|M - \lambda I| = 0$.

Multiplying both sides of the last equation by $|I-B|$ and utilizing the theorem concerning the product of the determinants of two matrices, we transform the equation into the form

$$|C-(I-B)\lambda| = 0,$$

or, in extended form,

(14)
$$\begin{vmatrix} a_{11}-\lambda & a_{12} & \cdots & a_{1n} \\ a_{21}\lambda & a_{22}-\lambda & \cdots & a_{2n} \\ \cdots & \cdots & \cdots & \cdots \\ a_{n1}\lambda & a_{n2}\lambda & \cdots & a_{nn}-\lambda \end{vmatrix} = 0.$$

Thus for convergence of the Seidel method, it is necessary and sufficient that all roots of the equation (14) be of modulus less than unity.

We shall now offer examples exhibiting the difference between the regions of convergence of Seidel's method and the ordinary method of iteration.

Example 1. Let

$$A = \begin{pmatrix} 5 & -5 \\ 1 & 0.1 \end{pmatrix}.$$

The proper numbers of the matrix A will then be determined from the equation $(0.1-\lambda)(5-\lambda)+5=0$, and therefore $|\lambda_i|>1$. The ordinary process of iteration diverges.

Let us form the matrix

$$(I-B)^{-1}C = \begin{pmatrix} 5 & -5 \\ 5 & -4.9 \end{pmatrix}.$$

The proper numbers of this matrix are determined from the equation

$$\lambda^2 - 0.1\lambda + 0.5 = 0.$$

Obviously $|\lambda_i| < 1$. The Seidel process converges.

Example 2. Let

$$A = \begin{pmatrix} 2.3 & -5 \\ 1 & -2.3 \end{pmatrix}.$$

The proper numbers of the matrix A are determined from the equation

$$-(2.3-\lambda)(2.3+\lambda)+5 = \lambda^2-0.29 = 0; \quad |\lambda_i| < 1.$$

The ordinary process of iteration converges.

It is readily verified that in this case the Seidel method diverges, for indeed, $(I-B)^{-1}C = \begin{pmatrix} 2.3 & -5 \\ 2.3 & -7.3 \end{pmatrix}$, and the proper numbers of this matrix are of modulus greater than unity.

If the system of equations is given in the form

(15) $$MX = G,$$

then in order to apply the Seidel method this system must be subjected to preparation to make it over into the form

$$X = AX+F.$$

The method most used leads to a process parallel to the process used for simple iteration, viz.: the system $MX=G$ is tabulated in the form

$$m_{ii}x_i = -\sum_{j=1}^{i-1} m_{ij}x_j - \sum_{j=i+1}^{n} m_{ij}x_j + g_i$$

or, what is the same thing, in the form

$$x_i = -\sum_{j=1}^{i-1} \frac{m_{ij}}{m_{ii}} x_j - \sum_{j=i+1}^{n} \frac{m_{ij}}{m_{ii}} x_j + \frac{g_i}{m_{ii}}$$

and the successive approximations are determined by the formulas

(16) $$x_i^{(k)} = -\sum_{j=1}^{i-1} \frac{m_{ij}}{m_{ii}} x_j^{(k)} - \sum_{j=i+1}^{n} \frac{m_{ij}}{m_{ii}} x_j^{(k-1)} + \frac{g_i}{m_{ii}}.$$

From the necessary and sufficient conditions of convergence in the general case considered above, those for the convergence of this modification of the process may be readily obtained, to wit: for the convergence of the process it is necessary and sufficient that all proper numbers of the matrix $Q^{-1}R$ be less than unity in modulus. Here

(17)

$$Q = \begin{pmatrix} m_{11} & 0 & \dots & 0 \\ m_{21} & m_{22} & \dots & 0 \\ \cdot & \cdot & \cdot & \cdot \\ m_{n1} & m_{n2} & \dots & m_{nn} \end{pmatrix}; \quad R = \begin{pmatrix} 0 & m_{12} & \dots & m_{1n} \\ 0 & 0 & \dots & m_{2n} \\ \cdot & \cdot & \cdot & \cdot \\ 0 & 0 & \dots & 0 \end{pmatrix}.$$

Indeed, the chosen preparation by which the system $MX = G$ is transformed into $X = AX + F$ is equivalent to a premultiplication of the system by a diagonal matrix P^{-1}, where P is the matrix composed of the elements of the principal diagonal of matrix M. The matrix $P^{-1}Q$ thereby plays the role of the matrix $(I-B)$, and the matrix $-P^{-1}R$ the role of the matrix C, and consequently $(I-B)^{-1}C = -Q^{-1}R$; the proper numbers of the matrices $-Q^{-1}R$ and $Q^{-1}R$ differ only in sign.

Furthermore, the characteristic polynomial $|-Q^{-1}R - \lambda I|$ of the matrix $-Q^{-1}R$, after multiplication by $|-Q|$, acquires the form $|R + \lambda Q|$. Consequently for the convergence of the specified modification of the Seidel process it is necessary and sufficient that all roots of the equation

(18)

$$\begin{vmatrix} m_{11}\lambda & m_{12} & \dots & m_{1n} \\ m_{21}\lambda & m_{22}\lambda & \dots & m_{2n} \\ \cdot & \cdot & \cdot & \cdot \\ m_{n1}\lambda & m_{n2}\lambda & \dots & m_{nn}\lambda \end{vmatrix} = 0$$

be of modulus less than one.

A great number of sufficient criteria of the convergence of this modification of the Seidel method is given in a list by P. A. Nekrasov and R. Mehmke.[1]

In particular, criteria I, II, I′, II′ of § 18 are sufficient.

In case the matrix M of the coefficients of the system $MX = G$ is symmetric, there exists still another important sufficient condition

[1] R. Mehmke and P. A. Nekrasov [1], P. A. Nekrasov [1].

of convergence of the modification of Seidel's process under consideration, viz.:

If the quadratic form with matrix M is positive-definite, the Seidel method for the system $MX = G$ converges.   This condition, on the assumption that the diagonal elements of the matrix M are positive, proves to be necessary as well.[1]

For the proof we shall put

$$(19) \qquad M = P + Q + Q',$$

where P is the diagonal matrix composed of the diagonal elements of the matrix M; Q is a triangular matrix formed by the elements of the matrix M that are located below the principal diagonal; and Q' is the matrix of the elements of M located above the principal diagonal. On the strength of the symmetry of M, Q' is the transpose of the matrix Q.

As we have seen above, the necessary and sufficient condition of convergence of the Seidel method is the requirement that all proper numbers of the matrix $T = (P + Q)^{-1}Q'$ be of modulus less than one.

Let μ_i and μ_j be any two proper numbers of the matrix T (they may, generally speaking, be complex), and let Z_i and Z_j be the proper vectors belonging to them.   Then

$$(P + Q)^{-1}Q'Z_i = \mu_i Z_i,$$

whence

$$(20) \qquad Q'Z_i = \mu_i P Z_i + \mu_i Q Z_i.$$

Analogously

$$Q'Z_j = \mu_j P Z_j + \mu_j Q Z_j.$$

We shall establish a relation of dependence between the scalar products (MZ_i, Z_j) and (PZ_i, Z_j), which will be important to all the subsequent discussion.

To this end let us express $(Q'Z_i, Z_j)$ and (QZ_i, Z_j) beforehand in terms of (PZ_i, Z_j).

We conclude from relation (20) that

$$(21) \qquad (Q'Z_i, Z_j) = \mu_i(PZ_i, Z_j) + \mu_i(QZ_i, Z_j).$$

[1] E. Reich [1].

Analogously

(22) $$(Q'Z_j, Z_i) = \mu_j(PZ_j, Z_i) + \mu_j(QZ_j, Z_i).$$

But, on the basis of the properties of the scalar product:

$$(Q'Z_j, Z_i) = (Z_j, QZ_i) = \overline{(QZ_i, Z_j)}$$

(23) $$(QZ_j, Z_i) = (Z_j, Q'Z_i) = \overline{(Q'Z_i, Z_j)}$$

$$(PZ_j, Z_i) = (Z_j, PZ_i) = \overline{(PZ_i, Z_j)}.$$

Using these relations, let us write equation (22) in the form

$$\overline{(QZ_i, Z_j)} = \mu_j\overline{(PZ_i, Z_j)} + \mu_j\overline{(Q'Z_i, Z_j)}$$

and, passing to the complex-conjugates,

(24) $$(QZ_i, Z_j) = \bar{\mu}_j(PZ_i, Z_j) + \bar{\mu}_j(Q'Z_i, Z_j).$$

From (21) and (24) we find

$$(QZ_i, Z_j) = \frac{\bar{\mu}_j(1+\mu_i)}{1-\mu_i\bar{\mu}_j} (PZ_i, Z_j)$$

and

$$(Q'Z_i, Z_j) = \frac{\mu_i(1+\bar{\mu}_j)}{1-\mu_i\bar{\mu}_j} (PZ_i, Z_j).$$

We have, moreover,

$$(MZ_i, Z_j) = (PZ_i, Z_j) + (QZ_i, Z_j) + (Q'Z_i, Z_j)$$

(25) $$= (1 + \frac{\bar{\mu}_j(1+\mu_i)}{1-\mu_i\bar{\mu}_j} + \frac{\mu_i(1+\bar{\mu}_j)}{1-\mu_i\bar{\mu}_j}) (PZ_i, Z_j)$$

$$= \frac{(1+\mu_i)(1+\bar{\mu}_j)}{1-\mu_i\bar{\mu}_j} (PZ_i, Z_j).$$

From the last relation the sufficiency of the formulated condition for the convergence of Seidel's method follows directly.

Indeed, putting $i=j$, we obtain

$$1 - \mu_i\bar{\mu}_i = \frac{(1+\mu_i)(1+\bar{\mu}_i)(PZ_i, Z_i)}{(MZ_i, Z_i)}.$$

But

$$\mu_i \bar{\mu}_i = |\mu_i|^2; \quad (1+\mu_i)(1+\bar{\mu}_i) = |1+\mu_i|^2.$$

Thus

$$1 - |\mu_i|^2 = \frac{|1+\mu_i|^2 (PZ_i, Z_i)}{(MZ_i, Z_i)} > 0,$$

for in accordance with the property of a positive-definite quadratic form, $(MZ_i, Z_i) > 0$ (see § 3, Paragraph 8), just as $(PZ_i, Z_i) > 0$. The last inequality follows from the fact that the diagonal terms of a positive-definite quadratic form themselves constitute a positive-definite quadratic form.

Proof of the necessity of the condition is somewhat more complex and is based on the following lemma from the theory of Hermitian forms.

Lemma. If

$$\Phi(x_1, x_2, \ldots, x_n) = \sum_{i, j = 1}^{n} c_{ij} x_i \bar{x}_j$$

is a positive-definite Hermitian form, $\alpha_1, \alpha_2, \ldots, \alpha_n$ are arbitrary complex numbers and $\mu_1, \mu_2, \ldots, \mu_n$ complex numbers less than of unit modulus, then

$$(26) \qquad \Psi(x_1, x_2, \ldots, x_n) = \sum_{i, j = 1}^{n} c_{ij} \frac{\alpha_i \bar{\alpha}_j}{1 - \mu_i \bar{\mu}_j} x_i \bar{x}_j$$

is also a positive-definite Hermitian form.

Proof. Indeed, on the strength of the fact that all $|\mu_i| < 1$,

$$\frac{1}{1 - \mu_i \bar{\mu}_j} = \sum_{k=0}^{\infty} \mu_i^k \bar{\mu}_j^k,$$

whence

$$\Psi = \sum_{k=0}^{\infty} \sum_{i, j = 1}^{n} c_{ij} \alpha_i \mu_i^k \bar{\alpha}_j \bar{\mu}_j^k x_i \bar{x}_j$$

$$= \sum_{k=0}^{\infty} \Phi(\alpha_1 \mu_1^k x_1, \alpha_2 \mu_2^k x_2, \ldots, \alpha_n \mu_n^k x_n) > 0,$$

for all terms of the series in question are positive.

The lemma is proved.

Let us return to the proof of the necessity of the formulated condition of convergence of the Seidel method, on the assumption of the positiveness of the diagonal elements of the matrix M.

Let the Seidel process converge. Then all proper numbers μ_i of the matrix $(P+Q)^{-1}Q'$, are of modulus less than unity. Let us assume, in addition, that they are all distinct, pair by pair. On this assumption the proper vectors $Z_1, Z_2, \ldots, Z_n$ are linearly independent and may be taken as a basis of the space.

Let X be any vector of the space and let $x_1, \ldots, x_n$ be its co-ordinates on the basis $Z_1, Z_2, \ldots, Z_n$, that is

$$X = x_1 Z_1 + \cdots + x_n Z_n.$$

We now introduce into the discussion the quadratic form $\Psi = (MX, X)$. In terms of the chosen coordinates it is expressible as an Hermitian form:

$$(27) \quad \Psi = \sum_{i,j=1}^{n} x_i \bar{x}_j (MZ_i, Z_i) = \sum_{i,j=1}^{n} \frac{(1+\mu_i)(1+\bar{\mu}_j)}{1-\mu_i \bar{\mu}_j} c_{ij} x_i \bar{x}_j$$

(on the basis of equation (25), where $c_{ij} = (PZ_i, Z_j)$).

The form

$$\Phi = \sum_{i,j=1}^{n} c_{ij} x_i \bar{x}_j = (PX, X)$$

is positive-definite, on the strength of the assumption of the positiveness of all elements of the diagonal matrix P.

Accordingly, on the strength of the lemma, the form Ψ is also positive-definite, which is what was required to be proved.

The assumption that was made regarding the absence of multiple proper numbers of the matrix T is not essential, for its satisfaction can be accomplished by an arbitrarily small deformation of the elements of the matrix M. Therefore from considerations of continuity, the positiveness of the form (MX, X) turns out to be a necessary condition of the convergence of the Seidel method (on the assumption that the diagonal elements of the matrix M are positive, of course) even given that multiple proper numbers are present.

We remark in conclusion that the method known as the "*relaxation method*" for the solution of linear systems is nothing more than an alteration of the computational scheme of the Seidel method. On this subject we refer the reader to an article by M. V. Nikolayeva.[1]

As an example let us find the solution of system (14) in § 17, having brought it into the form

$$x_1 = 0.22x_1 + 0.02x_2 + 0.12x_3 + 0.14x_4 + 0.76$$

$$x_2 = 0.02x_1 + 0.14x_2 + 0.04x_3 - 0.06x_4 + 0.08$$

$$x_3 = 0.12x_1 + 0.04x_2 + 0.28x_3 + 0.08x_4 + 1.12$$

$$x_4 = 0.14x_1 - 0.06x_2 + 0.08x_3 + 0.26x_4 + 0.68.$$

The sufficient conditions for the convergence of the Seidel process are obviously fulfilled.

TABLE XVI. *Computation of the Solution of the System by Seidel's Method*

$F = X^{(0)}$	0.76	0.08	1.12	0.68
$X^{(1)}$	1.1584	0.1184	1.6317	1.1424
$X^{(2)}$	1.3730	0.1208	1.8379	1.3090
$X^{(3)}$	1.4683	0.1213	1.9204	1.3723
$X^{(4)}$	1.5080	0.1216	1.9533	1.3969
$X^{(5)}$	1.5242	0.1218	1.9665	1.4066
$X^{(6)}$	1.5307	0.1219	1.9717	1.4104
$X^{(7)}$	1.5333	0.1220	1.9738	1.4120
$X^{(8)}$	1.5343	0.1220	1.9746	1.4126
$X^{(9)}$	1.53470	0.12200	1.97494	1.41281
$X^{(10)}$	1.53486	0.12201	1.97507	1.41290
$X^{(11)}$	1.53492	0.12201	1.97512	1.41293
$X^{(12)}$	1.534947	0.122009	1.975142	1.412945
$X^{(13)}$	1.5349579	0.1220094	1.9751507	1.4129513
$X^{(14)}$	1.5349622	0.1220095	1.9751541	1.4129538

[1] M. V. Nikolayeva [1].

As the initial approximation we will take the vector of constant terms. The successive approximations are set down in Table XVI.

Comparing the successive approximations that we have found with the solution of the system as found by the single-division method (see § 17), we see that the fourteenth approximation gives the solution to an accuracy of three units in the sixth place. The Seidel process, in the example before us, converges more rapidly than the ordinary process of iteration (see Table XIII, XIV and XV § 17).

§ 20. COMPARISON OF THE METHODS

In concluding this chapter we will make a comparison of the methods described, as regards their application to the problem of finding the solution of a non-homogeneous linear system.

The generally accepted criterion of the "advantageousness" of a given computational scheme is the number of necessary computational operations, that is, the number of necessary multiplications and divisions, and also additions and subtractions. Fundamental to this is the number of multiplications and divisions, and therefore in characterizing a method from this point of view we have limited ourselves to reckoning this number only. These criteria may not, however, be the only ones. Simplicity and uniformity of the operations to be performed in conformity with a given scheme may compel the computor to prefer it to a scheme with a considerably lesser number of operations but of more complicated "pattern". The latter circumstance is particularly important in case the computor has a digital calculator at his disposal.

Furthermore, one must take into consideration the possibility of a compact recording of the intermediate results, i.e., the possibility of carrying through the computations by accumulation formulas. Compact schemes, although they require higher qualifications on the part of the computor, frequently much simplify the entire process of computation. An important factor influencing the choice of a computational scheme is the absence of a loss of significant figures in the process of computation by a given scheme. Finally, the decisive factor is the reliability of the results obtained.

Let us analyse the methods we have described from the point of view of the factors indicated.

As was remarked earlier, the iterative methods offer greater simplicity in respect of the structure of their computation schemes. The separate operations are also carried out by accumulative means, and the computation can be carried through by a self-correcting process. Here it is easy to determine the number of operations necessary to obtain one iteration; to determine, however, the number k of necessary iterations is ordinarily possible only in the course of the work; it may prove to be very large.

Therefore in case the form of the system's coefficient matrix permits one to have confidence that the convergence of the iterative process will be sufficiently rapid (say $k < n$, where n is the order of the matrix of the system), this method may well be preferred to all others. Together with it one should without fail, of course, utilize the device that we analyse in § 35 for improving the convergence of the process.

Systems, however, exist for which the generally adopted iterative methods either converge too slowly or even diverge. In this case only the exact methods make possible the computation of the solution.

Modifications of the most powerful of the exact methods—the Gauss method—are the compact schemes, those of the type of the single-division scheme, and the elimination scheme. The compact schemes of the Gauss method, and in particular the square-root method for systems with symmetric matrix, can be recommended in case highly qualified computors are available. The single-division scheme may sometimes be successfully rivalled by the elimination scheme, in which the whole process of computations is conducted uniformly, thanks to the absence of a return course.

The number of operations by the different variants of the Gauss method is approximately the same; only the number of results to be recorded are different. As has already been remarked in describing the Gauss method, its basic defect is the possible disappearance of significant figures in case either the determinant of the system itself or that of one of the intermediate minors is small.

In such cases the escalator method is reliable (we have described it only for symmetric systems).

Finally, the determination of the solution by means of the inverse matrix, theoretically the simplest, is in practice sometimes also the most efficient method.

This will be the case, for example, when the solution must be determined with a high degree of accuracy. Indeed, as we saw in § 13, the elements of the inverse matrix can be determined as exactly as one wishes, as soon as one has found a sufficiently exact approximation, from which an iterative process giving these elements converges extremely rapidly. For computing the first approximation to the inverse matrix, along with the method of Gauss, the method of bordering may also be employed with success.

CHAPTER III

THE PROPER NUMBERS AND PROPER VECTORS OF A MATRIX

This chapter is devoted to the problems of finding the proper numbers and the proper vectors of a matrix.   We recall that what are spoken of as the proper numbers of a matrix A are the zeros of its characteristic polynomial, i.e., the roots of the equation

$$|A - \lambda I| = \begin{vmatrix} a_{11} - \lambda & a_{12} & \cdots & a_{1n} \\ a_{21} & a_{22} - \lambda & \cdots & a_{2n} \\ \cdots \cdots \cdots \cdots \cdots \cdots \\ a_{n1} & a_{n2} & \cdots & a_{nn} - \lambda \end{vmatrix}$$

$$= (-1)^n [\lambda^n - p_1 \lambda^{n-1} - \cdots - p_n] = 0.$$

As has already been remarked in § 1, the coefficients p_i are, but for sign, the sums of all principal ith order minors of the determinant of the matrix A.   The direct computation of the coefficients p_i is extremely awkward and requires a huge number of operations.

The determination of the components of a proper vector requires the solution of a system of n homogeneous equations in n unknowns; in order to compute all the vectors of a matrix, one must solve, generally speaking, n systems of the form

$$(A - \lambda_i I) X_i = 0,$$

where $X_i = (x_{1i}, x_{2i}, \ldots, x_{ni})$ is the ith proper vector of the matrix A.

It is thus perfectly natural that special computational artifices that simplify the numerical solution of both the problems before us should have made their appearance. As in the preceding chapter, we shall distinguish two groups of methods: exact and iterative. The exact methods, when applied to the first problem, will give more or less convenient schemes for determining the coefficients p_i. The proper numbers will then be obtained as the solutions of an algebraic equation of the nth degree. We note that all of the methods proposed below, with the exception of that of Leverrier (1840), have appeared in the thirties of our century or later. The majority of the methods proposed are adapted to the solution of the first problem. We shall show, however, that by utilizing the intermediate results of the computations, the proper vectors of the matrix belonging to the corresponding proper numbers can also be computed after the latter (avoiding the solution of the systems indicated above). Only one method—namely that known as the escalator method—solves both problems simultaneously, but it makes necessary the computation, in the course of the process, of the complete spectrum of the matrix, and all the proper vectors not only of the given matrix but of its transpose as well.[1]

The iterative methods make possible the direct determination of the proper numbers of the matrix, without resorting to the characteristic polynomial. In using them, only the first proper numbers are as a rule determined with sufficient accuracy. The course of the iterative process depends essentially on the structure of the Jordan canonical form that is connected with the given matrix and on whether the proper numbers of the matrix are real or imaginary.

---

[1] We note that the determination of the roots of the characteristic equation may be efficiently carried out by Newton's method, viz.: if

$$\varphi(\lambda) = \lambda^n - p_1\lambda^{n-1} - \cdots - p_n = 0$$

is the characteristic equation, then $\lambda_1 = \lambda_0 - \dfrac{\varphi(\lambda_0)}{\varphi'(\lambda_0)}$, where λ_0 is some initial approximation to a root, and λ_1 gives a significantly better approximation than λ_0. The values of $\varphi(\lambda_0)$ are conveniently calculated by Horner's method: $\varphi(\lambda_0) = q_n$, where the numbers q_k $(k=0, \ldots, n)$ are computed by the recurrence formulas

$$q_k = q_{k-1}\lambda_0 - p_k, \quad q_{-1} = 0.$$

Analogously

$$\varphi'(\lambda_0) = r_{n-1}, \quad \text{where } r_k = r_{k-1}\lambda_0 + q_k, \quad r_{-1} = 0.$$

The convergence of the iterative process is determined by the magnitude of the ratio of the moduli of different, neighboring proper numbers, and may be very slow. We shall exhibit several artifices making it possible to much improve the convergence of the process. A proper vector is computed simultaneously with the proper number to which it belongs. As in the solution of linear systems, the chief merit of the iterative processes consists in the simplicity and uniformity of the operations to be performed.

In all the following sections we shall assume the coefficients of the matrix to be *real*.

§ 21. THE METHOD OF A. N. KRYLOV

A. N. Krylov's work [1] was the first in a great cycle of works devoted to bringing the secular equation into polynomial form.

The idea of A. N. Krylov consists in making a preliminary transformation of the equation

$$(1) \qquad D(\lambda) = \begin{vmatrix} a_{11}-\lambda & a_{12} & \cdots & a_{1n} \\ a_{21} & a_{22}-\lambda & \cdots & a_{2n} \\ \cdot & \cdot \cdot \cdot \cdot \cdot \cdot \cdot \cdot & \cdot \\ a_{n1} & a_{n2} & \cdots & a_{nn}-\lambda \end{vmatrix} = 0$$

into an equation of the form

$$(2) \qquad D_n(\lambda) = \begin{vmatrix} b_{11}-\lambda & b_{12} & \cdots & b_{1n} \\ b_{21}-\lambda^2 & b_{22} & \cdots & b_{2n} \\ \cdot \cdot \cdot \cdot & \cdot \cdot \cdot & \cdot \\ b_{n1}-\lambda^n & b_{n2} & \cdots & b_{nn} \end{vmatrix} = 0.$$

which is, generally speaking, equivalent to it, and whose expansion in powers of λ is obviously accomplished considerably more simply by expanding this determinant by minors of the first column.

In order to effect the transformation indicated, A. N. Krylov introduces into the discussion a differential equation, which is connected with the given matrix; he simultaneously poses the problem of finding a purely algebraic transformation turning equation (1) into equation (2).

Works of N. N. Luzin [1], [2], I. N. Khlodovsky [1], F. R. Gantmakher [1], and D. K. Faddeev [2] are devoted to explaining the algebraic essence of the transformation of A. N. Krylov's.

We shall present A. N. Krylov's method in an algebraic interpretation.

That the determinant

$$
(3) \qquad D(\lambda) = \begin{vmatrix} a_{11}-\lambda & a_{12} & \cdots & a_{1n} \\ a_{21} & a_{22}-\lambda & \cdots & a_{2n} \\ \cdots\cdots\cdots\cdots\cdots\cdots \\ a_{n1} & a_{n2} & \cdots & a_{nn}-\lambda \end{vmatrix}
$$

equals zero is the necessary and sufficient condition that the system of homogeneous equations

$$
(4) \qquad \begin{aligned} \lambda x_1 &= a_{11}x_1 + a_{12}x_2 + \cdots + a_{1n}x_n \\ \lambda x_2 &= a_{21}x_1 + a_{22}x_2 + \cdots + a_{2n}x_n \\ &\cdots\cdots\cdots\cdots\cdots\cdots\cdots \\ \lambda x_n &= a_{n1}x_1 + a_{n2}x_2 + \cdots + a_{nn}x_n \end{aligned}
$$

have a solution $x_1, x_2, \ldots, x_n$ different from zero.

Let us transform system (4) in the following manner. We shall multiply the first equation by λ and replace $\lambda x_1, \ldots, \lambda x_n$ by their expressions (4) in terms of $x_1, \ldots, x_n$.

This gives

$$
(5) \qquad \lambda^2 x_1 = b_{21}x_1 + b_{22}x_2 + \cdots + b_{2n}x_n,
$$

where

$$
(6) \qquad b_{2k} = \sum_{s=1}^{n} a_{1s}a_{sk}.
$$

Moreover, let us multiply equation (5) by λ and again replace $\lambda x_1, \lambda x_2, \ldots, \lambda x_n$ by their expressions in terms of $x_1, \ldots, x_n$. We shall obtain

$$
\lambda^3 x_1 = b_{31}x_1 + b_{32}x_2 + \cdots + b_{3n}x_n.
$$

Repeating this process $(n-1)$ times, we pass from system (4) to the system

(7)
$$\lambda x_1 = b_{11}x_1 + b_{12}x_2 + \cdots + b_{1n}x_n$$
$$\lambda^2 x_1 = b_{21}x_1 + b_{22}x_2 + \cdots + b_{2n}x_n$$
$$\cdot \ \cdot \ \cdot \ \cdot \ \cdot \ \cdot \ \cdot \ \cdot \ \cdot \ \cdot \ \cdot \ \cdot \ \cdot$$
$$\lambda^n x_1 = b_{n1}x_1 + b_{n2}x_2 + \cdots + b_{nn}x_n,$$

whose coefficients b_{ik} will be determined by the recurrence formulas

(8)
$$b_{ik} = \sum_{s=1}^{n} b_{i-1,\,s}a_{sk} \quad \begin{array}{l} i=2,\ldots,n \\ k=1,\ldots,n \end{array}$$
$$b_{1k} = a_{1k}.$$

The determinant of system (7) will obviously have form (2).

System (7) has a non-zero solution for all values of λ satisfying the equation $D(\lambda)=0$. Thus $D_n(\lambda)$ reduces to zero for all λ that are roots of the equation $D(\lambda)=0$.

We shall show that

(9)
$$\frac{D_n(\lambda)}{D(\lambda)} = \begin{vmatrix} 1 & 0 & \ldots & 0 \\ b_{11} & b_{12} & \ldots & b_{1n} \\ \cdot & \cdot & \cdot & \cdot \\ b_{n-1,\,1} & b_{n-1,\,2} & \ldots & b_{n-1,\,n} \end{vmatrix} = N,$$

that is, for $N \neq 0$, $D_n(\lambda)$ differs from the sought characteristic polynomial by a numerical factor only.

Let all roots of $D(\lambda)$ be distinct. Since all roots of $D(\lambda)$ are roots of $D_n(\lambda)$, $D_n(\lambda)$ is divisible by $D(\lambda)$. Since, in addition, the degrees of $D_n(\lambda)$ and $D(\lambda)$ are alike, the quotient must be a constant (does not depend on λ). Comparing the coefficients of λ^n, we obtain

$$\frac{D_n(\lambda)}{D(\lambda)} = N.$$

In case $D(\lambda)$ has multiple roots, the equation

(10)
$$D_n(\lambda) = ND(\lambda)$$

is preserved, which follows, however, from considerations of continuity.

This equation may also be verified directly by a multiplication of the determinants figuring in it if relations (8) are used.

It is evident from equation (10) that if $N=0$, $D_n(\lambda)$ is identically equal to zero. In this case the indicated transformation yields nothing.

However even for $N=0$, A. N. Krylov proposes a special method whose algebraic substance will be elucidated below.

Let us now turn to the coefficients b_{ik} defining $D_n(\lambda)$. We introduce into the discussion the vectors B_i with components $(b_{i1}, b_{i2}, \ldots, b_{in})$. The equations

$$(8) \qquad \begin{aligned} b_{ik} &= \sum_{s=1}^{n} b_{i-1,\,s} a_{sk} \quad &i=2,\ldots,n \\ b_{1k} &= a_{1k} &k=1,\ldots,n, \end{aligned}$$

show that

$$(11) \qquad B_i = A'B_{i-1},$$

A' being the transpose of the given matrix.

From equation (11) it follows that

$$B_i = A'^{i-1}B_1, \quad i=2,\ldots,n.$$

In its turn $B_1 = A'B_0$, where $B_0 = (1, 0, \ldots, 0)$. Thus, finally,

$$(12) \qquad B_i = A'^i B_0, \quad i=1, 2, \ldots, n.$$

It is obvious that system (4) may be transformed by starting, for example, from the second equation of that system. In this case λ will be put into the second column of the determinant $D_n(\lambda)$, and the coefficients b_{ik} will be determined by formulas (12), where $B_0 = (0, 1, 0, \ldots, 0)$.

The method of A. N. Krylov is quite naturally generalized if we introduce into the discussion instead of the vector B_0 of special form, the arbitrary vector $B_0 = (b_{01}, b_{02}, \ldots, b_{0n})$.

Let

$$(13) \qquad u = b_{01}x_1 + b_{02}x_2 + \cdots + b_{0n}x_n,$$

where $(x_1, x_2, \ldots, x_n)$ is the solution of system (4).

Then on repeating the preceding argument, we will obtain

$$u = b_{01}x_1 + b_{02}x_2 + \cdots + b_{0n}x_n$$
$$\lambda u = b_{11}x_1 + b_{12}x_2 + \cdots + b_{1n}x_n$$
(14)
$$\lambda^2 u = b_{21}x_1 + b_{22}x_2 + \cdots + b_{2n}x_n$$
$$\cdot \quad \cdot \quad \cdot \quad \cdot \quad \cdot \quad \cdot \quad \cdot \quad \cdot \quad \cdot \quad \cdot \quad \cdot \quad \cdot \quad \cdot$$
$$\lambda^n u = b_{n1}x_1 + b_{n2}x_2 + \cdots + b_{nn}x_n,$$

where $B_i = (b_{i1}, b_{i2}, \ldots, b_{in}) = A'^i B_0$.

Regarding the $n+1$ equations (14) as a system of homogeneous linear equations in $n+1$ unknowns $u, x_1, \ldots, x_n$, we will have a non-zero solution possible in case, and only in case, the determinant

(15)
$$D_n(\lambda) = \begin{vmatrix} 1 & b_{01} & \ldots & b_{0n} \\ \lambda & b_{11} & \ldots & b_{1n} \\ \cdot & \cdot & \cdot & \cdot \\ \lambda^n & b_{n1} & \ldots & b_{nn} \end{vmatrix} = 0.$$

Repeating the previous reasoning, we find that

$$D_n(\lambda) = D(\lambda)N,$$

where this time

(16)
$$N = \begin{vmatrix} b_{01} & b_{02} & \ldots & b_{0n} \\ b_{11} & b_{12} & \ldots & b_{1n} \\ \cdot & \cdot & \cdot & \cdot \\ b_{n-1, 1} & b_{n-1, 2} & \ldots & b_{n-1, n} \end{vmatrix}.$$

Just as for the particular case considered above, the transformation yields nothing if $N=0$.

Let us assume at the outset, therefore, that $N \neq 0$. On the basis of the equation $D_n(\lambda) = ND(\lambda)$, the coefficients p_i of the characteristic polynomial are to be determined as the ratios $\dfrac{(-1)^{n-1}N_i}{N}$, where N_i are the algebraic complements of the elements λ^{n-i} in the determinant $D_n(\lambda)$. And the determination of the coefficients of the characteristic polynomial by means of the ratios indicated indeed constitutes the essence of A. N. Krylov's work. However, the investigation there conducted makes possible the determination of the sought-for coefficients while avoiding the computation of the minors, thus much reducing the number of operations necessary.

Indeed, in view of the fact that the rows of determinant (16) are the components of the vectors $B_0, B_1, \ldots, B_{n-1}$, the condition $N \neq 0$ is equivalent to the linear independence of these vectors. Therefore for $N \neq 0$, the vectors $B_0, B_1, \ldots, B_{n-1}$ form a basis of the space. Accordingly the vector B_n is a linear combination of them:

$$(17) \qquad B_n = q_1 B_{n-1} + \cdots + q_n B_0.$$

We shall show that the coefficients of this dependence are indeed the coefficients p_i of the characteristic polynomial, the latter being written in the form

$$D(\lambda) = (-1)^n [\lambda^n - p_1 \lambda^{n-1} - \cdots - p_n].$$

Indeed, taking from the last row of the determinant $D_n(\lambda)$ a linear combination of the preceding rows with respective coefficients $q_1, q_2, \ldots, q_n$, we obtain, on the basis of equation (17),

$$D_n(\lambda) = \begin{vmatrix} 1 & & b_{01} & \cdots & b_{0n} \\ \cdots & \cdots & \cdots & \cdots & \cdots \\ \lambda^{n-1} & & b_{n-1,\,1} & \cdots & b_{n-1,\,n} \\ \lambda^n - q_1\lambda^{n-1} - \cdots - q_n & & 0 & \cdots & 0 \end{vmatrix}$$

$$= (-1)^n [\lambda^n - q_1 \lambda^{n-1} - \cdots - q_n] N.$$

Hence

$$D(\lambda) = \frac{D_n(\lambda)}{N} = (-1)^n [\lambda^n - q_1 \lambda^{n-1} - \cdots - q_n],$$

which is what was required to be proved.

Equation (17) permits the finding of the coefficients $q_1 = p_1$, $q_2 = p_2, \ldots, q_n = p_n$ as the solutions of a system of linear equations equivalent to this vector equation.

Equation (17) connects the method of A. N. Krylov with the Cayley-Hamilton relation (applied to the matrix A').

Indeed, from the relation

$$A'^n = p_1 A'^{n-1} + \cdots + p_n I$$

it follows that

$$A'^n B_0 = p_1 A'^{n-1} B_0 + \cdots + p_n B_0,$$

that is, that

$$(17) \qquad B_n = p_1 B_{n-1} + \cdots + p_n B_0.$$

In place of system (17), one can obviously employ for the determination of the coefficients p_i the system

$$(17') \qquad C_n = p_1 C_{n-1} + \cdots + p_n C_0,$$

where the vectors C_n are determined by the equations $C_k = A^k C_0$.

For the determination of the coefficients p_i by solving system (17) or (17'), one must carry out $\frac{3}{2} n^2 (n+1)$ multiplications and divisions. In its original form the method of A. N. Krylov required $\frac{1}{3}(n^4 + 4n^3 + 2n^2 - n - 3)$ multiplications and divisions.

In case $N = 0$, the system equivalent to equation (17) does not make possible the determination of the coefficients of the characteristic polynomial, since the determinant of this system is likewise equal to N.

The algebraic essence of the device of A. N. Krylov's alluded to above consists in this: it is possible to determine the coefficients of the polynomial of least degree, $\Psi(\lambda)$, such that $\Psi(A)C_0 = 0$. Generally speaking, this will be the minimum polynomial of the matrix, and its roots will coincide with all the roots of the characteristic polynomial, but will be of lower multiplicity. However, with an unfortunate choice of the vector C_0, instead of the minimum polynomial some one of its divisors may be obtained, whereupon part of the roots of the equation $|A - \lambda I| = 0$ may be lost. As has been shown by N. N. Luzin and I. N. Khlodovsky,[1] with a special choice of the vector C_0 one can obtain as the polynomial $\Psi(\lambda)$ any divisor of the minimum polynomial.

We note that if the minimum polynomial of the matrix does not coincide with the characteristic polynomial, $N = 0$ for any choice of

---

[1] N. N. Luzin [1], [2]; I. N. Khlodovsky [1]; F. R. Gantmakher [1]; D. K. Faddeev [2].

the vector C_0, for $\Psi(A)C_0 = 0$, and since the degree of the polynomial $\Psi(\lambda)$ is less than n, the vectors $C_0, AC_0, \ldots, A^{n-1}C_0$ are linearly dependent. As regards an additional degeneration, this may be avoided by changing the initial vector C_0.

Thus the method of A. N. Krylov makes possible the determination of the coefficients of the characteristic polynomial if $N \neq 0$, or of some one of its divisors—generally speaking the minimum polynomial—if $N = 0$.

In practice the circumstance that $N = 0$ will reveal itself during the forward course of the solution of system (17) by the Gauss method. For in part of the equations all coefficients will be simultaneously eliminated, so that these equations will reduce to the identity $0 = 0$. These equations (let them be $n - m$ in number) must be discarded; in the system that remains it will be necessary to discard $n - m$ of the last columns, beginning with the column of constant terms (i.e., with the components of the vector C_n). The last of the remaining columns, composed of the components of the vector C_m, must be taken as the constant member of the new system. The solution of the system gives the coefficients of the linear dependence of C_m on $C_0, C_1, \ldots, C_{m-1}$, that is, the coefficients of some divisor only of the characteristic polynomial.

We shall here analyse two examples, both taken from the article of A. N. Krylov's [1].

As the first example let us determine the coefficients of the characteristic polynomial of the matrix

$$\begin{pmatrix} -5.509882 & 1.870086 & 0.422908 & 0.008814 \\ 0.287865 & -11.811654 & 5.711900 & 0.058717 \\ 0.049099 & 4.308033 & -12.970687 & 0.229326 \\ 0.006235 & 0.269851 & 1.397369 & -17.596207 \end{pmatrix},$$

which A. N. Krylov took from Leverrier's work [1]. By Leverrier's computation,

$$\varphi(\lambda) = \lambda^4 + 47.888430\lambda^3 + 797.2789\lambda^2 + 5349.457\lambda + 12296.555 = 0;$$

$$\lambda_1 = -17.86303; \quad \lambda_2 = -17.15266; \quad \lambda_3 = -7.57404,$$

$$\lambda^4 = -5.29870.$$

In Table I we exhibit the scheme for the determination of the coefficients p_i, regarding them as solutions of system (17').

There, in rows 1 to 4, are arrayed the components of the vectors $A^k C_0$ $(k = 0, 1, 2, 3, 4)$ as they are computed, and also the usual check sums. (For computational convenience we tabulate the matrix A on a separate sheet in transposed form, and obtain the components of the vector being computed as the sum of the products of the elements of the corresponding column of the tabulated matrix and the components of the preceding vector, in one accumulation.) The fifth row effects the check, which is analogous to that employed in the example of § 17.

TABLE I. *Computation of the Coefficients of the Characteristic Equation by the Method of A. N. Krylov. Leverrier's Matrix*

C_0	C_1	C_2	C_3	C_4	Σ
1	−5.509882	30.917951	−179.01251	1100.7201	948.11566
0	0.287865	−4.705449	66.38829	−967.5973	−905.62659
0	0.049099	0.334184	−23.08728	576.5226	553.81860
0	0.006235	0.002224	−0.649152	−4.04003	−4.68072
	−5.166683	26.548910	−136.3606 (7)	705.6054	
	1	−16.34603	230.62300	−3361.2885	−3146.0115
		1.136758	−34.41064	741.5585	708.28462
		0.104141	−2.087086	16.91760	14.93466
		1	−30.27086	652.3451	623.0742
			1.065352	−51.01827	−49.95291 (2)
			1	−47.8887	−46.8886 (7)
		1		−797.287	−796.284 (7)
	1			−5349.53	−5348.50 (3)
1				−12296.8	−12295.7 (8)

In rows 6 to 14 is contained the solution of the system obtained; we find this solution by the single-division scheme.

As a final check on the computation of the coefficients p_i we have a comparison of the value of p_1 with the trace of the matrix. Since tr $A = -47.888430$, we see that the value of p_1 found from the solution of the system is sufficiently exact.

A comparison with Leverrier's data shows, however, that the coefficients have been computed with a lesser degree of accuracy. This loss of accuracy is well-known, and is the inherent shortcoming of the A. N. Krylov method; it is to be explained by the circumstance that the coefficients of the system determining the p_i are quantities of different orders of magnitude.

A somewhat better result can be attained by employing the pivotal condensation scheme for the solution of the system.

As the second example we will take yet another matrix from the article of A. N. Krylov's [1]:

$$\begin{pmatrix} 5 & 30 & -48 \\ 3 & 14 & -24 \\ 3 & 15 & -25 \end{pmatrix}$$

In the subjoined table the first three rows contain the coefficients of the system for the determination of the p_i:

1	5	-29	125
0	3	-15	63
0	3	-15	63
	0	0	0

The first step of the Gauss process shows that here we have a case of degeneration. The truncated system

$$q_1 + 5q_2 = -29$$
$$3q_2 = -15$$

gives as the solution $q_1 = -5$, $q_2 = -4$. Thus in this case we have determined the coefficients of the second-degree polynomial $\lambda^2 + 5\lambda + 4$, which is but a divisor of the characteristic polynomial.

§ 22. THE DETERMINATION OF PROPER VECTORS BY THE METHOD OF A. N. KRYLOV

Let us assume that the coefficients of the characteristic polynomial have been determined by the method of A. N. Krylov, and that all the proper numbers have been calculated and have proved to be distinct. We shall show how to determine the proper vectors of the matrix by utilizing computations already carried out. Let C_0 be the initial vector in the A. N. Krylov process, and let $X_1, X_2, \ldots, X_n$ be the proper vectors of the matrix A belonging to $\lambda_1, \lambda_2, \ldots, \lambda_n$. In § 3 it has been shown that $X_1, \ldots, X_n$ are linearly independent.

Let us resolve the vector C_0 in terms of the proper vectors:

$$(1) \qquad C_0 = \alpha_1 X_1 + \alpha_2 X_2 + \cdots + \alpha_n X_n.$$

We have, moreover,

$$C_1 = AC_0 = \alpha_1 \lambda_1 X_1 + \alpha_2 \lambda_2 X_2 + \cdots + \alpha_n \lambda_n X_n$$

$$(2) \qquad \cdot \quad \cdot \quad \cdot \quad \cdot \quad \cdot \quad \cdot \quad \cdot \quad \cdot \quad \cdot \quad \cdot \quad \cdot$$

$$C_{n-1} = A^{n-1} C_0 = \alpha_1 \lambda_1^{n-1} X_1 + \alpha_2 \lambda_2^{n-1} X_2 + \cdots + \alpha_n \lambda_n^{n-1} X_n.$$

The vectors $C_1, C_2, \ldots, C_{n-1}$ have been computed in the course of finding the proper numbers. Let us form a linear combination of them:

$$
\begin{aligned}
\beta_{10} C_{n-1} &+ \beta_{11} C_{n-2} + \cdots + \beta_{1,\, n-1} C_0 \\
&= \alpha_1 (\beta_{10} \lambda_1^{n-1} + \beta_{11} \lambda_1^{n-2} + \cdots + \beta_{1,\, n-1}) X_1 \\
&\quad + \alpha_2 (\beta_{10} \lambda_2^{n-1} + \beta_{11} \lambda_2^{n-2} + \cdots + \beta_{1,\, n-1}) X_2 \\
&\quad + \cdot \quad \cdot \quad \cdot \quad \cdot \quad \cdot \quad \cdot \quad \cdot \quad \cdot \quad \cdot \quad \cdot \\
&\quad + \alpha_n (\beta_{10} \lambda_n^{n-1} + \beta_{11} \lambda_n^{n-2} + \cdots + \beta_{1,\, n-1}) X_n \\
&= \alpha_1 \varphi_1(\lambda_1) X_1 + \alpha_2 \varphi_1(\lambda_2) X_2 + \cdots + \alpha_n \varphi_1(\lambda_n) X_n,
\end{aligned}
$$

(3)

where

$$(4) \qquad \varphi_1(\lambda) = \beta_{10} \lambda^{n-1} + \beta_{11} \lambda^{n-2} + \cdots + \beta_{1,\, n-1}.$$

Let us select the coefficients $\beta_{10}, \ldots, \beta_{1, n-1}$ so that

(5) $\varphi_1(\lambda_1) \neq 0, \quad \varphi_1(\lambda_2) = \cdots = \varphi_1(\lambda_n) = 0.$

To achieve this it is sufficient to take as $\varphi_1(\lambda)$ the polynomial

$$\varphi_1(\lambda) = (\lambda - \lambda_2) \ldots (\lambda - \lambda_n)$$

(6)
$$= \frac{(\lambda - \lambda_1)(\lambda - \lambda_2) \ldots (\lambda - \lambda_n)}{\lambda - \lambda_1} = \frac{(-1)^n \varphi(\lambda)}{\lambda - \lambda_1}$$

$$= \frac{(-1)^n(\lambda^n - p_1\lambda^{n-1} - \cdots - p_n)}{\lambda - \lambda_1}.$$

Here $\varphi(\lambda)$ is the characteristic polynomial, whose coefficients and roots are already computed.

The coefficients of the quotient (6) can easily be computed by the recurrence formulas

(7) $\beta_{10} = 1, \quad \beta_{1j} = \lambda_1 \beta_{1, j-1} - p_j,$
$$j = 1, \ldots, n-1 \text{ (Horner's scheme)},$$

which issue directly from the equation

$$(\lambda - \lambda_1)(\beta_{10}\lambda^{n-1} + \beta_{11}\lambda^{n-2} + \cdots + \beta_{1, n-1})$$
$$= (-1)^n[\lambda^n - p_1\lambda^{n-1} - \cdots - p_n].$$

Thus
$$\beta_{10}C_{n-1} + \beta_{11}C_{n-2} + \cdots + \beta_{1, n-1}C_0 = \alpha_1\varphi_1(\lambda_1)X_1,$$

that is to say, the linear combination that we formed is the proper vector X_1, accurate to a numerical factor. Of course the coefficient α_1 must be different from zero; this is guaranteed by the successful completion of the A. N. Krylov process. Since the proper vector is determinate but for a constant factor, we may adopt for the proper vector the constructed linear combination.

Analogously

(8) $X_i = \sum_{j=0}^{n-1} \beta_{ij}C_{n-1-j},$

where

(9) $\beta_{i0} = 1, \quad \beta_{ij} = \lambda_i\beta_{i, j-1} - p_j, \quad j = 1, \ldots, n-1.$

As an example we shall compute, for the Leverrier matrix, the proper vector belonging to the proper number $\lambda_4 = -5.29870$.

We reproduce the characteristic equation with the coefficients taken from Table *I*:

$$\lambda^4 + 47.8887\lambda^3 + 797.287\lambda^2 + 5349.53\lambda + 12296.8 = 0.$$

Computing the numbers β_{4j} for $j = 0, 1, 2, 3$, we obtain

$$1; \quad 42.5900; \quad 571.615; \quad 2320.71.$$

We form the linear combinations in accordance with formula (8), arranging the computations in the table

$\beta_{43}C_0$	$\beta_{42}C_1$	$\beta_{41}C_2$	$\beta_{40}C_3$	X_4	$\tilde{X}_4$
2320.71	−3149.53	1316.80	−179.01	308.97	1
0	164.548	−200.405	66.388	30.531	0.098815
0	28.066	14.233	−23.087	19.212	0.062181
0	3.5640	0.0947	−0.6492	3.0095	0.009740

The last column contains the components of the proper vector X_4, normalized so that its first component equals unity. We shall see below that the values of the components of the proper vector computed by the method of A. N. Krylov agree well with the values computed by other methods.

§ 23. SAMUELSON'S METHOD

The method suggested by P. Samuelson [1] is close to the A. N. Krylov method.

The computational scheme of this method is as follows. One computes the rectangular matrix

(1)

$$\begin{pmatrix} R & \vdots & 0 & 0 & 0 & 0 & \ldots & 0 & 0 & 1 & -a_{11} \\ RM & \vdots & 0 & 0 & 0 & 0 & \ldots & 0 & 1 & -a_{11} & -RS \\ RM^2 & \vdots & 0 & 0 & 0 & 0 & \ldots & 1 & -a_{11} & -RS & -RMS \\ \cdots & \vdots & \cdots & \cdots & \cdots & \cdots & & & & & \\ RM^{n-1} & \vdots & 1 & -a_{11} & -RS & -RMS & \ldots & & & & -RM^{n-2}S \end{pmatrix},$$

where R, S, and M are cells in the following partition of the given matrix:

$$(2) \quad A = \begin{pmatrix} a_{11} & \vdots & a_{11} & \cdots & a_{1n} \\ \cdots\cdots & \vdots & \cdots\cdots\cdots\cdots\cdots \\ a_{21} & \vdots & a_{22} & \cdots & a_{2n} \\ \cdot\ \cdot\ \cdot & \vdots & \cdot\ \cdot\ \cdot\ \cdot\ \cdot\ \cdot\ \cdot \\ a_{n1} & \vdots & a_{n2} & \cdots & a_{nn} \end{pmatrix} = \begin{pmatrix} a_{11} & R \\ S & M \end{pmatrix}.$$

Furthermore, by means of elementary transformations (as this is done in the problem of elimination, § 12) one must attain this situation: that in place of the row RM^{n-1} a zero row turns up. Then the remaining elements of the last row will give, generally speaking, the coefficients of the characteristic polynomial. The process of elimination, as we have seen, is very uniform and simple. This is the fundamental merit of the scheme.

The author derives his scheme from a transformation of a system of linear differential equations connected with the matrix into one differential equation of order n, by means of a special device for elimination. A brief algebraic rationale of the scheme is contained in the following.

Let

$$(3) \qquad X_0 = \begin{pmatrix} x_{10} \\ x_{20} \\ \cdot\ \cdot \\ x_{n0} \end{pmatrix} = \begin{pmatrix} x_{10} \\ Y_0 \end{pmatrix}$$

be an arbitrary vector.

Furthermore, let

$$AX_0 = \begin{pmatrix} x_{11} \\ x_{21} \\ \vdots \\ x_{n1} \end{pmatrix} = \begin{pmatrix} x_{11} \\ Y_1 \end{pmatrix}, \quad \ldots, A^{n-1}X_0 = \begin{pmatrix} x_{1,\,n-1} \\ x_{2,\,n-1} \\ \vdots \\ x_{n,\,n-1} \end{pmatrix} = \begin{pmatrix} x_{1,\,n-1} \\ Y_{n-1} \end{pmatrix},$$

$$(4)$$

$$A^n X_0 = \begin{pmatrix} x_{1n} \\ x_{2n} \\ \vdots \\ x_{nn} \end{pmatrix} = \begin{pmatrix} x_{1n} \\ Y_n \end{pmatrix}.$$

From the construction it follows that

(5) $$x_{1k} = a_{11}x_{1, k-1} + RY_{k-1}$$

(6) $$Y_k = Sx_{1, k-1} + MY_{k-1} \quad k = 1, \ldots, n.$$

Thus we have n^2 relations (n and $n(n-1)$) between $n^2 + n$ quantities. They make possible the elimination from the system of equations (5) and (6) of the vectors $Y_1, \ldots, Y_n$, that is, of $n(n-1)$ quantities. As the result of this elimination there remain n equations connecting $2n$ numbers, these being the components of the vector Y_0 and the numbers $x_{10}, \ldots, x_{1n}$.

Let us carry out this elimination. We have, for $k = 1, 2, \ldots, n$:

$$\begin{aligned}
x_{1k} &= a_{11}x_{1, k-1} + RY_{k-1} \\
&= a_{11}x_{1, k-1} + RSx_{1, k-2} + RMY_{k-2} \\
&= a_{11}x_{1, k-1} + RSx_{1, k-2} + RMSx_{1, k-3} + RM^2Y_{k-3} \\
&= \ldots \ldots \ldots \ldots \ldots \ldots \ldots \ldots \\
&= a_{11}x_{1, k-1} + RSx_{1, k-2} + RMSx_{1, k-3} \\
&\qquad\qquad + \cdots + RM^{k-2}Sx_{10} + RM^{k-1}Y_0,
\end{aligned}$$

or

(7) $$RM^{k-1}Y_0 = x_{1k} - a_{11}x_{1, k-1} - RSx_{1, k-2} - \cdots - RM^{k-2}Sx_{10}.$$

The coefficients of these n equations obviously form matrix (1).

Eliminating from these n equations the components of the vector Y_0, we shall obtain a single linear relation between the numbers $x_{10}, \ldots, x_{1n}$ with constant coefficients independent of the choice of the initial vector.

On the other hand, proceeding from the Cayley-Hamilton relation, we have

$$x_{1n} - p_1 x_{1, n-1} - \cdots - p_n x_{10} = 0.$$

This equation is also a relation of linear dependence between the numbers $x_{10}, \ldots, x_{1n}$, with constant coefficients independent of the choice of the vector.

This relation of dependence will coincide with the relation of dependence obtained by the method of elimination in case the matrix A is such that we may rightly consider the numbers $x_{10}, \ldots, x_{1, \, n-1}$ to be independent variables, that is, in case we may assign them, independently of one another, arbitrary values by using a suitable selection of the remaining components of the initial vector X_0, or, in other words, by using the vector Y_0.

The Samuelson method may be more rigorously proved by means of the following relations between the coefficients of the characteristic polynomials of the outer and inner matrices of our bordered system.

Let

$$
\begin{aligned}
f(\lambda) &= (-1)^n[\lambda^n + p_1\lambda^{n-1} + \cdots + p_n] \\
\varphi(\lambda) &= (-1)^n[\lambda^{n-1} + q_1\lambda^{n-2} + \cdots + q_{n-1}]
\end{aligned}
$$
(8)

be the characteristic polynomials of the matrices A and M. (Contravening the usual notational usage for the polynomials, we have changed the signs of the coefficients p_k and q_k.)

Then the following relations are valid:

$$
\begin{aligned}
p_1 &= -a_{11} + q_1 \\
p_2 &= -RS - q_1 a_{11} + q_2 \\
p_3 &= -RMS - q_1 RS - q_2 a_{11} + q_3 \\
&\cdot \quad \cdot \quad \cdot \quad \cdot \quad \cdot \quad \cdot \quad \cdot \quad \cdot \quad \cdot \quad \cdot \quad \cdot \\
p_{n-1} &= -RM^{n-3}S - q_1 RM^{n-4}S - \cdots + q_{n-1} \\
p_n &= -RM^{n-2}S - q_1 RM^{n-3}S - \cdots - q_{n-2}RS - q_{n-1}a_{11}.
\end{aligned}
$$
(9)

These relations are obtained from the rule for the expansion of a bordered determinant.

Moreover, if the A. N. Krylov method be applied to the matrix M, having adopted R' (with the components $a_{12}, \ldots, a_{1n}$) for the initial vector, the coefficients $q_1, q_2, \ldots, q_{n-1}$ will then be defined by the system of equations

$$
M'^{n-1}R' + q_1 M'^{n-2}R' + \cdots + q_{n-1}R' = 0.
$$
(10)

TABLE II. *Computation of the Coefficients of the Characteristic Equation by Samuelson's Scheme. Leverrier's Matrix*

								Σ
1.870086	0.422908	0.008814	0	0	0	1	5.509882	8.811690
-20.264529	5.208653	0.051697	0	0	1	5.509882	-0.559152	-9.053449
261.81060	-183.23653	-0.905064	0	1	5.509882	-0.559152	5.577387	89.197123
-3882.0495	3870.8749	-10.72247	1	5.509882	-0.559152	5.577387	-66.36373	-76.73268
1	0.226144	0.004713	0	0	0	0.534735	2.946325	4.711917
	9.791355	0.147204	0	0	1	16.346035	59.146736	86.431330
	-242.44343	-2.138977	0	1	5.509882	-140.55844	-765.80173	-1144.4327
	4748.777	7.57363	1	5.509882	-0.559152	2081.445	11371.416	18215.162
	1	0.015034	0	0	0.102131	1.669435	6.040710	8.827310
		1.505918	0	1	30.270872	264.18511	698.72872	995.6906
		-63.81948	1	5.509882	-485.5565	-5846.330	-17314.569	-23703.765 (6)
		1	0	0.664047	20.10128	175.43127	463.98856	661.18514
			1	47.88902	797.297	5349.60	12296.94	18492.73 (3)

On the strength of relations (9), the coefficients $p_1, p_2, \ldots, p_n$ are non-homogeneous linear forms in $q_1, \ldots, q_{n-1}$ and consequently may be calculated simultaneously by the method of elimination (see § 12). Of the two possible modifications of the method of elimination, one should take that in which the components of the vectors $R', M'R', \ldots, M'^{n-1}R'$, are arrayed in the rows of the scheme. Then these rows, conceived as matrices, are $R, RM, \ldots, RM^{n-1}$. The coefficients of the relations (9) will thereupon turn out to be arranged so as to accord exactly with the Samuelson scheme. Given the grounds for the method that have been presented above, the region of its application is easily discerned. Indeed, it coincides with the region of application of the A. N. Krylov method for the matrix M, proceeding from the vector R'.

As an example we shall again take the Leverrier matrix. We carry through the computation of the coefficients of the characteristic polynomial in accordance with the scheme described (see Table II). At the outset we compute matrix (1), arranging its elements in the first four rows. We next perform the elimination as was shown in § 12. The last row gives the sought values of the coefficients, which as regards accuracy almost coincide with the values computed by the method of A. N. Krylov. The last column, as usual, is the check column.

The number of operations necessary to determine the coefficients of the characteristic polynomial by the Samuelson method is somewhat less than by the A. N. Krylov method, for the formation of matrix (1) requires $n(n-1)^2$ multiplications, and the process of elimination in the Samuelson scheme requires just as many operations as does the solution of the system in the A. N. Krylov method. In connection with this we comment that in Wayland's article [1] the reckoning of the number of operations for the Samuelson method is incorrectly done.

§ 24. THE METHOD OF A. M. DANILEVSKY

An elegant and very efficient method of computing the coefficients of the characteristic polynomial has been proposed by A. M. Danilevsky [1]. The gist of his method consists in an initial reduction of the secular determinant to the form known as the Frobenius normal form:

$$D(\lambda) = \begin{vmatrix} p_1 - \lambda & p_2 & p_3 & \dots & p_n \\ 1 & -\lambda & 0 & \dots & 0 \\ 0 & 1 & -\lambda & \dots & 0 \\ \cdot & \cdot & \cdot & \cdot & \cdot \\ 0 & 0 & 0 & \dots & -\lambda \end{vmatrix},$$

the expansion of which in powers of λ demands no labor, since

$$D(\lambda) = (-1)^n[\lambda^n - p_1\lambda^{n-1} - \cdots - p^n].$$

On the strength of the fact that similar matrices have identical characteristic polynomials, in order to attain the aim set, it is sufficient to reduce the given matrix

$$A = \begin{pmatrix} a_{11} & a_{12} & \dots & a_{1n} \\ a_{21} & a_{22} & \dots & a_{2n} \\ \cdot & \cdot & \cdot & \cdot \\ a_{n1} & a_{n2} & \dots & a_{nn} \end{pmatrix}$$

to the form

$$P = \begin{pmatrix} p_1 & p_2 & \dots & p_{n-1} & p_n \\ 1 & 0 & \dots & 0 & 0 \\ 0 & 1 & \dots & 0 & 0 \\ \cdot & \cdot & \cdot & \cdot & \cdot \\ 0 & 0 & \dots & 1 & 0 \end{pmatrix}$$

by means of a similarity transformation.

We shall show that we can find $(n-1)$ similarity transformations, the successive performance of which will realize the desired transition from matrix A to matrix P, if this is possible.

Let us examine the beginning of the process.

We must carry the row $a_{n1}\ a_{n2}\ \dots\ a_{n,\,n-1}\ a_{nn}$ into the row $0\ 0\ \dots\ 1\ 0$. Let us assume that $a_{n,\,n-1} \neq 0$. Divide all elements of the row by $a_{n,\,n-1}$; then subtract the $(n-1)$th column, multiplied by $a_{n1}, a_{n2}, \dots, a_{nn}$ respectively, from all the rest of the columns. The desired transformation will thereby obviously have been effected.

The operations we have indicated will thus be elementary transformations on columns, and, as has been shown in § 1 Paragraph 10, will reduce to a postmultiplication of the matrix A by a matrix M_{n-1} which is not hard to jot down, to wit:

$$
M_{n-1} = \begin{pmatrix}
1 & 0 & \ldots & 0 & 0 \\
0 & 1 & \ldots & 0 & 0 \\
\multicolumn{5}{c}{\cdot\ \cdot\ \cdot\ \cdot\ \cdot\ \cdot\ \cdot\ \cdot\ \cdot\ \cdot\ \cdot\ \cdot\ \cdot\ \cdot} \\
-\dfrac{a_{n1}}{a_{n,\,n-1}} & -\dfrac{a_{n2}}{a_{n,\,n-1}} & \ldots & \dfrac{1}{a_{n,\,n-1}} & -\dfrac{a_{nn}}{a_{n,\,n-1}} \\
0 & 0 & \ldots & 0 & 1
\end{pmatrix}.
$$

The constructed matrix AM_{n-1} will not be similar to the matrix A. However from it one can easily pass to a matrix that is similar to A. To accomplish this it is enough to premultiply the matrix AM_{n-1} by the matrix M_{n-1}^{-1}, which is easily computed; indeed one can directly verify that

$$
M_{n-1}^{-1} = \begin{pmatrix}
1 & 0 & \ldots & 0 & 0 \\
0 & 1 & \ldots & 0 & 0 \\
\multicolumn{5}{c}{\cdot\ \cdot\ \cdot\ \cdot\ \cdot\ \cdot\ \cdot\ \cdot\ \cdot\ \cdot\ \cdot\ \cdot} \\
a_{n1} & a_{n2} & \ldots & a_{n,\,n-1} & a_{nn} \\
0 & 0 & \ldots & 0 & 1
\end{pmatrix}.
$$

The multiplication of the matrix AM_{n-1} on the left by the matrix M_{n-1}^{-1} obviously does not change the transformed row.

Thus the matrix

$$
M_{n-1}^{-1}AM_{n-1} = \begin{pmatrix}
c_{11} & c_{12} & \ldots & c_{1,\,n-1} & c_{1n} \\
c_{21} & c_{22} & \ldots & c_{2,\,n-1} & c_{2n} \\
\multicolumn{5}{c}{\cdot\ \cdot\ \cdot\ \cdot\ \cdot\ \cdot\ \cdot\ \cdot\ \cdot\ \cdot\ \cdot\ \cdot\ \cdot\ \cdot} \\
c_{n-1,\,1} & c_{n-1,\,2} & \ldots & c_{n-1,\,n-1} & c_{n-1,\,n} \\
0 & 0 & \ldots & 1 & 0
\end{pmatrix} = C
$$

has one satisfactory row, and we can continue the process another step further if $c_{n-1,\,n-2} \neq 0$. If all the $(n-1)$ transformations are possible, we shall have brought the matrix A into form P.

An explanation of how one is to proceed in those exceptional cases when the necessary elements vanish will be given below.

At the moment we shall show how the elements of the matrix C, the result of one similarity transformation, are computed.

We have

$$AM_{n-1} = \begin{pmatrix} a_{11} & a_{12} & \cdots & a_{1,\,n-1} & a_{1n} \\ \cdots\cdots\cdots\cdots\cdots\cdots\cdots\cdots\cdots\cdots \\ a_{n-1,\,1} & a_{n-1,\,2} & \cdots & a_{n-1,\,n-1} & a_{n-1,\,n} \\ a_{n1} & a_{n2} & \cdots & a_{n,\,n-1} & a_{nn} \end{pmatrix}$$

$$\times \begin{pmatrix} 1 & 0 & \cdots & 0 & 0 \\ \cdots\cdots\cdots\cdots\cdots\cdots\cdots\cdots\cdots\cdots \\ m_{n-1,\,1} & m_{n-1,\,2} & \cdots & m_{n-1,\,n-1} & m_{n-1,\,n} \\ 0 & 0 & \cdots & 0 & 1 \end{pmatrix}$$

$$= \begin{pmatrix} b_{11} & b_{12} & \cdots & b_{1,\,n-1} & b_{1n} \\ b_{21} & b_{22} & \cdots & b_{2,\,n-1} & b_{2n} \\ \cdots\cdots\cdots\cdots\cdots\cdots\cdots\cdots\cdots \\ b_{n-1,\,1} & b_{n-1,\,2} & \cdots & b_{n-1,\,n-1} & b_{n-1,\,n} \\ 0 & 0 & \cdots & 1 & 0 \end{pmatrix} = B,$$

where

$$b_{ik} = a_{ik} + a_{i,\,n-1}m_{n-1,\,k} \qquad \text{for all } i \leqslant n-1, \quad k \neq n-1,$$
$$b_{i,\,n-1} = a_{i,\,n-1}m_{n-1,\,n-1}.$$

Here

$$m_{n-1,\,i} = -\frac{a_{ni}}{a_{n,\,n-1}} \qquad \text{for } i \neq n-1$$

and

$$m_{n-1,\,n-1} = \frac{1}{a_{n,\,n-1}}.$$

We see that all the elements of the matrix $B = AM_{n-1}$, with the exception of the elements of the $(n-1)$th column, are to be computed by formulas involving two terms.

The elements of the $(n-1)$th column, if we recall the value of the coefficients $m_{n-1,\,n-1}$, are obtained as the quotients of the division of

TABLE III. *Computational Scheme for the Method of A. M. Danilevsky*

	I	1	2	3	4	Σ	Σ'
1		−5.509882	1.870086	0.422908	0.008814	−3.208074	
2		0.287865	−11.811654	5.711900	0.058717	−5.753172	
3		0.049099	4.308033	−12.970687	0.229326	−8.384229	
4		0.006235	0.269851	1.397369	−17.596207	−15.922752	
4'	$M_3^{-1} \downarrow \vec{M_3}$	−0.004462	−0.193114	0.71563059 / −1	12.592384	11.394808	
5	0.006235	−5.511769	1.788417	0.302646	5.334234	1.913527 (8)	1.610881
6	0.269851	0.262379	−12.914702	4.087610	71.985155	63.420442	59.332832
7	1.397369	0.106974	6.812854	9.282220	−163.10255	−165.46494	−156.18272
8	−17.596207	0	0	1	0	1	
9		0.185920	6.046177	−29.461961	−208.45592	−231.68578	
9'	$M_2^{-1} \downarrow \vec{M_2}$	−0.030750	0.16539377 / −1	4.872825	34.47731	38.31938	
10	0.185920	−5.566763	0.295793	9.017289	66.99404	70.74035 (6)(90)	70.44456
11	6.046177	0.659506	−2.136011	−58.84347	−373.2790	−433.5989	−431.4629
12	−29.461961	0	1	0	0	1	
13	−208.45592	0	0	1	0	1	
14		2.952517	−42.32167	−562.5575	−2244.455	−2846.382 (6)	
14'	$M_1^{-1} \downarrow \vec{M_1}$	0.33869407 / −1	14.33410	190.5349	760.1836	964.0527 (6)	
15	2.952517	−1.885430	−79.49874	−1051.645	−4164.768	−5297.798 (7)	−5295.913
16	−42.32167	1	0	0	0	1	
17	−562.5575	0	1	0	0	1	
18	−2244.455	0	0	1	0	1	
19		−47.88843	−797.2789	−5349.455	−12296.55	−18491.17	

the elements of the $(n-1)$th column of the matrix A by the element $a_{n,\,n-1}$.

Moreover,

$$
C = M_{n-1}^{-1} A M_{n-1} = \begin{pmatrix} 1 & 0 & \cdots & 0 & 0 \\ 0 & 1 & \cdots & 0 & 0 \\ \cdot & \cdot & \cdot & \cdot & \cdot \\ a_{n1} & a_{n2} & \cdots & a_{n,\,n-1} & a_{nn} \\ 0 & 0 & \cdots & 0 & 1 \end{pmatrix}
$$

$$
\times \begin{pmatrix} b_{11} & b_{12} & \cdots & b_{1,\,n-1} & b_{1n} \\ b_{21} & b_{22} & \cdots & b_{2,\,n-1} & b_{2n} \\ \cdot & \cdot & \cdot & \cdot & \cdot \\ b_{n-1,\,1} & b_{n-1,\,2} & \cdots & b_{n-1,\,n-1} & b_{n-1,\,n} \\ 0 & 0 & \cdots & 1 & 0 \end{pmatrix}
$$

$$
= \begin{pmatrix} b_{11} & b_{12} & \cdots & b_{1,\,n-1} & b_{1n} \\ b_{21} & b_{22} & \cdots & b_{2,\,n-1} & b_{2n} \\ \cdot & \cdot & \cdot & \cdot & \cdot \\ \sum_{k=1}^{n} a_{nk} b_{k1} & \sum_{k=1}^{n} a_{nk} b_{k2} & \cdots & \sum_{k=1}^{n} a_{nk} b_{k,\,n-1} & \sum_{k=1}^{n} a_{nk} b_{kn} \\ 0 & 0 & \cdots & 1 & 0 \end{pmatrix}.
$$

Thus the premultiplication by M_{n-1}^{-1} changes only the $(n-1)$th row of the matrix B; the elements of this row are found with one accumulation. The entire process, consisting of $(n-1)$ similarity transformations by means of the matrices $M_{n-1}, M_{n-2}, \ldots, M_1$, is fitted into a convenient computational scheme, which we shall show in an example.

As the example we again take the Leverrier matrix (Table III). The Danilevsky transformation consists in the successive multiplication of the given matrix on the right by the matrices M_3, M_2, and M_1, and on the left by the matrices M_3^{-1}, M_2^{-1}, M_1^{-1}, where the right and left multiplications alternate. Each matrix of the transformation is given as one row.

Let us pass on to the description of the computational process. In rows 1, 2, 3, 4 are arrayed the elements of the given matrix and the check sums. We begin the computation with the computation of the elements defining the matrices M_3 and M_3^{-1}, viz.: in row 4′ we enter the elements of the third row of the matrix M_3:

$$m_{31} = -\frac{a_{41}}{a_{43}}; \quad m_{32} = -\frac{a_{42}}{a_{43}}; \quad m_{33} = \frac{1}{a_{43}}; \quad m_{34} = -\frac{a_{44}}{a_{43}}.$$

The elements of the third row of the matrix M_3^{-1}, equal to the elements $a_{4k}(k=1, 2, 3, 4)$ we enter, for computational convenience, in a column: rows 5, 6, 7, 8 of the head-column I. The result of the multiplication by M_3 we write in the four rows 5, 6, 7, 8. The transformation M_3 brings the fourth row of the matrix into canonical form (the 8th row). The elements of the third column are now obtained by multiplying the elements of the third column of the matrix A by $m_{33} = \frac{1}{a_{43}}$. The rest of the elements are computed by the two-term formulas:

$$(1) \qquad\qquad b_{ik} = a_{ik} + a_{i3}m_{3k}.$$

For the check we form the column of sums, as usual. The element of the check column located in row 4′ must coincide with the sum of the elements of this row after replacing the element m_{33} by -1. The results of the application of formulas (1) to the elements $\sum$ are entered in the column $\sum'$; adding the elements of the 3rd column to them, we obtain the check sums for the rows 5 to 8.

The transformation M_3^{-1} changes only the 7th row; we enter the result in the 9th row. Its elements are obtained as the sum of the products, by pairs, of the elements situated in column I by the corresponding elements of each column of the matrix AM_3. With this we have concluded the transformation $M_3^{-1}AM_3$. The process is continued analogously. We remark that the elements of the 9th row are also the elements of the 2nd row of the matrix M_2^{-1}; we copy them columnwise in column I. The elements of the second row of the matrix M_2 are entered in row 9′.

The process is easily learned and is afterwards executed without difficulty.

As the result of the computation we obtain all the coefficients at once, since the coincidence—the check—of p_1 with the trace of the matrix is in addition an index to the accuracy of the computation of the rest of the coefficients; the results obtained are closer to the Leverrier data than are the results found by A. N. Krylov's method or those by the Samuelson method.

The number of operations necessary for computation by the A. M. Danilevsky method is substantially less than by the two other methods referred to above. The number of multiplications and divisions is equal to $(n-1)(n^2+n-1)$.

We remark that A. M. Danilevsky's method admits of a modification similar to the pivotal condensation scheme in connection with the Gauss method; it somewhat increases the accuracy of the results obtained.

Knowledge of the matrices $M_1, M_2, \ldots, M_{n-1}$ which successively effect the similarity transformations, permits us to determine the proper vectors of the matrix A.

Indeed, let λ be some proper number of matrix A and $X = (x_1, \ldots, x_n)$ the proper vector corresponding to it. Let $Y = (y_1, y_2, \ldots, y_n)$ be a proper vector of the matrix P. Then, as was shown in § 3 Paragraph 10,

$$X = M_{n-1}M_{n-2} \ldots M_2 M_1 Y.$$

Now, the proper vector of the matrix P is found without trouble, since its components will be the solutions of the recurrence system:

$$(p_1 - \lambda)y_1 + p_2 y_2 + \cdots + p_n y_n = 0$$

$$y_1 - \lambda y_2 \qquad\qquad\qquad = 0$$

$$\cdot \quad \cdot \quad \cdot \quad \cdot \quad \cdot \quad \cdot \quad \cdot \quad \cdot \quad \cdot \quad \cdot \quad \cdot \quad \cdot$$

$$y_{n-1} - \lambda y_n \qquad\qquad\qquad = 0.$$

This gives $y_n = 1, y_{n-1} = \lambda, \ldots, y_1 = \lambda^{n-1}$.

The transformation M_1, performed upon Y, gives

$$M_1 Y = \begin{pmatrix} m_{11} & m_{12} & \cdots & m_{1n} \\ 0 & 1 & \cdots & 0 \\ \cdot & \cdot & \cdots & \cdot \\ 0 & 0 & \cdots & 1 \end{pmatrix} \begin{pmatrix} y_1 \\ y_2 \\ \cdot \cdot \\ y_n \end{pmatrix} = \begin{pmatrix} \sum_{k=1}^{n} m_{1k} y_k \\ y_2 \\ \cdot \cdot \cdot \cdot \\ y_n \end{pmatrix}.$$

The transformation M_1 thus alters only the first component of the vector Y. Analogously, the transformation M_2 alters the second component, etc.

Thus the components of the vector X are determined by the formulas

$$(2) \qquad x_k = \sum_{s=1}^{k-1} m_{ks}x_s + \sum_{s=k}^{n} m_{ks}y_s \quad (k=1, \ldots n-1).$$

As an example let us consider, for the Leverrier matrix, the computation of the proper vector belonging to the proper number λ_4.

From the equation

$$\lambda^4 + 47.88843\lambda^3 + 797.2789\lambda^2 + 5349.455\lambda + 12296.55 = 0$$

we obtain $\lambda_4 = -5.29872$. The computation of the proper vector belonging to λ_4 is performed in conformity with the subjoined table.

TABLE IV.   *Computation of a Proper Vector by the
A. M. Danilevsky Method*

	I	II	III	IV	V	VI
1	0.338694	−0.030750	−0.004462	−148.769	102.655	1.000000
2	14.33410	0.165394	−0.193114	28.0764	10.1446	0.098822
3	190.5349	4.872825	0.715631	−5.29872	6.38334	0.062182
4	760.1836	34.47731	12.59238	1	1	0.009741

Here columns I, II, III contain the elements m_{1k}, m_{2k}, m_{3k}, which we transcribe from rows 14′, 9′ and 4′ of Table III, setting them up in columns. Column IV contains powers of λ_4, from the third to the zero-th. Column V contains the components of the proper vector, which are computed one after another by formula (2), which is reminiscent of the formula of the Seidel method. Column VI contains the components of the vector X_4, normalized so that the first of its components is equal to unity.

In concluding this section we shall dwell on exceptional cases.

Let us assume that after several steps of the process we have arrived at a matrix of the form

$$C = \begin{pmatrix} c_{11} & c_{12} & \cdots & c_{1k} & & \cdots & & c_{1n} \\ c_{21} & c_{22} & \cdots & c_{2k} & & \cdots & & c_{2n} \\ \cdot & \cdot & \cdot & \cdot & \cdot & \cdot & \cdot & \cdot \\ c_{k1} & c_{k2} & \cdots & c_{kk} & & \cdots & & c_{kn} \\ 0 & 0 & \cdots & 1 & 0 & \cdots & & 0 \\ 0 & 0 & \cdots & 0 & 1 & \cdots & & 0 \\ \cdot & \cdot & \cdot & \cdot & \cdot & \cdot & \cdot & \cdot \\ 0 & 0 & \cdots & 0 & 0 & \cdots & 1 & 0 \end{pmatrix}$$

and together with this it turns out that $c_{k,\,k-1} = 0$.

Here two cases are conceivable.

If any one of the elements c_{ki} $(i < k-1)$ is different from zero, we interchange the ith and $(k-1)$th columns and simultaneously change the rows with the same numbers. Such a transformation is equivalent to a multiplication left and right by a matrix S (see Paragraph 10, § 1) of the form

This matrix has the property that $S^2 = I$, and, accordingly,

$S = S^{-1}$, so that a multiplication on left and right by S is a similarity transformation. After these transformations one must proceed as usual.

If, however, all $c_{ki} = 0$ for $i \leqslant k-1$, then the matter becomes even simpler.

To wit, in this case the matrix C has the form

$$C = \begin{pmatrix} c_{11} & c_{12} & \cdots & c_{1,\,k-1} & c_{1,\,k} & \cdots & c_{1n} \\ \cdots & \cdots & \cdots & \cdots & \cdots & \cdots & \cdots \\ c_{k-1,\,1} & c_{k-1,\,2} & \cdots & c_{k-1,\,k-1} & c_{k-1,\,k} & \cdots & c_{k-1,\,n} \\ 0 & 0 & \cdots & 0 & c_{kk} & \cdots & c_{kn} \\ 0 & 0 & \cdots & 0 & 1 & \cdots & 0 \\ \cdots & \cdots & \cdots & \cdots & \cdots & \cdots & \cdots \\ 0 & 0 & \cdots & 0 & 0 & \cdots 1 & 0 \end{pmatrix} = \begin{pmatrix} C_1 & D \\ 0 & C_2 \end{pmatrix},$$

where

$$C_1 = \begin{pmatrix} c_{11} & \cdots & c_{1,\,k-1} \\ \cdots & \cdots & \cdots \\ c_{k-1,\,1} & \cdots & c_{k-1,\,k-1} \end{pmatrix}, \quad C_2 = \begin{pmatrix} c_{kk} & \cdots & c_{kn} \\ 1 & \cdots & 0 \\ \cdots & \cdots & \cdots \\ 0 & \cdots & 1 & 0 \end{pmatrix}$$

and consequently

$$|C - \lambda I| = |C_1 - \lambda I| \cdot |C_2 - \lambda I|.$$

The matrix C_2 already has the Frobenius canonical form and therefore $|C_2 - \lambda I|$ is computed instantly. In order to expand $|C_1 - \lambda I|$, one must apply the general process leading to the matrix P_1.

Thus the case where the process is broken introduces only simplifications into the problem of computing the characteristic polynomial: an easily computable factor is separated from the characteristic polynomial, and the remaining factor is the characteristic polynomial of a matrix of lower order.

§ 25. LEVERRIER'S METHOD IN D. K. FADDEEV'S MODIFICATION

In this section we shall expound a method known as Leverrier's method [2], requiring a greater number of operations than any of the methods presented above, but utterly insensitive to the individual peculiarities of the matrix, in particular to "gaps" in the intermediate determinants.

Let

$$(1) \qquad D_n(\lambda) = (-1)^n[\lambda^n - p_1\lambda^{n-1} - p_2\lambda^{n-2} - \cdots - p_n]$$

be the characteristic polynomial of the matrix and $\lambda_1, \lambda_2, \ldots, \lambda_n$ its roots, among which some may be equal. Let us employ the symbol

$$(2) \qquad \sum_{l=1}^{n} \lambda_l^k = s_k.$$

Then a relation is valid that is known as the Newton formula:

$$(3) \qquad kp_k = s_k - p_1 s_{k-1} - \cdots - p_{k-1}s_1, \quad k = 1, \ldots, n.$$

If the numbers s_k are known, then by solving the recurrence system (3) we can find the coefficients p_k which we need.

We shall show how the numbers s_k are determined. We have

$$s_1 = \lambda_1 + \lambda_2 + \cdots + \lambda_n = \operatorname{tr} A.$$

Moreover, on the strength of Paragraph 11, § 3, the characteristic numbers of the matrix A^k will be $\lambda_1^k, \lambda_2^k, \ldots, \lambda_n^k$. Accordingly

$$(4) \qquad s_k = \lambda_1^k + \lambda_2^k + \cdots + \lambda_n^k = \operatorname{tr} A^k.$$

Thus the process of computation reduces to the successive computation of the powers of the matrix A, then to the calculation of their traces and, finally, to the solution of the recurrence system (3). The computation of the n powers of the matrix A (of the last matrix A^n it is necessary to compute only the diagonal elements) requires a great number of operations—uniform, granted—and Leverrier's method is inordinately more laborious than the methods expounded above. Its value consists, as has already been mentioned, in its

TABLE V. D. K. Faddeev's Modification of Leverrier's Method

Determination of the Coefficients of the Characteristic Polynomial and Elements of the Inverse Matrix

(1)	(2)	(3)	(4)	(5)	(6)	(7)	(8)	Σ
−5.509882	0.287865	0.049099	0.006235	42.378548	1.870086	0.422908	0.008814	44.68036
1.870086	−11.811654	4.308033	0.269851	0.287865	36.076776	5.711900	0.058717	42.13526
0.422908	5.711900	−12.970687	1.397369	0.049099	4.308033	34.917743	0.229326	39.50420
0.008814	0.058717	0.229326	−17.596207	0.006235	0.269851	1.397369	30.292223	31.96568
$p_1 = -47.888430$								
−232.94165				564.33711	58.98700	23.13088	0.42522	646.8802
	−400.96516			9.07995	396.31360	132.18346	2.39755	539.9745 (6)
		−427.95884		2.68546	99.69550	369.31992	4.22567	475.9266
			−532.69187	0.30081	11.01857	25.74858	264.58689	301.6549 (8)
$p_2 = -797.27876$								
−3091.3122				2258.1433	458.3882	276.1613	6.2599	2998.953
	−4094.0411			70.5604	1255.4144	556.3836	11.4757	1893.835 (4)
		−4213.8419		32.0618	419.6360	1135.6136	16.2164	1603.527 (8)
			−4649.1712	4.4283	52.7398	98.8129	700.2843	856.264 (5)
$p_3 = -5349.4555$								
−12296.551	0.000	−0.000	−0.000	−0.183640	−0.037278	−0.022458	−0.000509	
−0.001	−12296.551	0.000	0.000	−0.005738	−0.102095	−0.045247	−0.000933	
0.001	0.000	−12296.550	0.001	−0.002607	−0.034126	−0.092352	−0.001319	
0.001	−0.001	0.002	−12296.551	−0.000360	−0.004289	−0.008036	−0.056950	
$p_4 = -12296.551$								

universality. The number of multiplications necessary in the Leverrier method is equal to $\frac{1}{2}(n-1)(2n^3 - 2n^2 + n + 2)$.

We remark that in computing the powers of a matrix it is useful to perform a check by means of a column composed of the sums of the elements of each row of the matrix A. The result of the multiplication of the matrix A by this column must coincide with an analogous column of the matrix A^2, for let $\sum_1$ be the column of the sums of matrix A, $\sum_2$ that of the sums of the matrix A^2, and let

$$U = \begin{pmatrix} 1 & & & \\ & 1 & & \\ & & \ddots & \\ & & & 1 \end{pmatrix}.$$

Then:

$$\sum_1 = AU, \quad \sum_2 = A^2 U,$$

that is,

$$(5) \qquad\qquad \sum_2 = A\sum_1.$$

What has been said is obviously true for the other powers too.

We now expound the modification of the method that was proposed by D. K. Faddeev[1] which, in addition to simplifying the computation of the coefficients of the characteristic polynomial, permits us to determine the inverse matrix and the proper vectors of the matrix.

Instead of the sequence $A, A^2, \ldots, A^n$, let us compute the sequence $A_1, A_2, \ldots, A_n$, constructed in the following manner:

$$A_1 = A, \qquad \operatorname{tr} A_1 = q_1, \quad B_1 = A_1 - q_1 I$$

$$A_2 = AB_1, \quad \frac{\operatorname{tr} A_2}{2} = q_2, \quad B_2 = A_2 - q_2 I$$

$$(6) \qquad \cdot \quad \cdot \quad \cdot \quad \cdot \quad \cdot \quad \cdot \quad \cdot \quad \cdot \quad \cdot \quad \cdot \quad \cdot \quad \cdot \quad \cdot \quad \cdot \quad \cdot$$

$$A_{n-1} = AB_{n-2}, \quad \frac{\operatorname{tr} A_{n-1}}{n-1} = q_{n-1}, \quad B_{n-1} = A_{n-1} - q_{n-1} I$$

$$A_n = AB_{n-1}, \qquad \frac{\operatorname{tr} A_n}{n} = q_n, \qquad B_n = A_n - q_n I.$$

[1] D. K. Faddeev and I. S. Sominsky [1].

We shall show that a) $q_1 = p_1, q_2 = p_2, \ldots, q_n = p_n;$

b) B_n is a null matrix;

c) if A is a nonsingular matrix, then

$$A^{-1} = \frac{B_{n-1}}{p_n}.$$

(If the matrix A is singular, then $(-1)^{n-1}B_{n-1}$ will be the adjoint of matrix A.)

We shall prove a) by the method of mathematical induction. That $p_1 = \text{tr } A = q_1$ is obvious. We shall assume that $q_1 = p_1, q_2 = p_2, \ldots, q_{k-1} = p_{k-1}$, and shall prove that $q_k = p_k$. In accordance with our construction:

$$A_k = A^k - q_1 A^{k-1} - \cdots - q_{k-1}A$$

$$= A^k - p_1 A^{k-1} - \cdots - p_{k-1}A.$$

Accordingly

$$\text{tr } A_k = kq_k = \text{tr } A^k - p_1 \text{tr } A^{k-1} - \cdots - p_{k-1}\text{tr } A$$

$$= s_k - p_1 s_{k-1} - \cdots - p_{k-1}s_1.$$

Hence, on the strength of the Newton formula, $kq_k = kp_k, q_k = p_k$. Furthermore, on the strength of the Cayley-Hamilton relation:

(7) $B_n = A^n - p_1 A^{n-1} - \cdots - p_n I = 0.$

Hence it follows that $A_n = p_n I$. This last fact may be utilized for checking the computation; it is obvious that the deviation of A_n from a scalar matrix is a measure of the accuracy of the computations. Besides this final check it is convenient to utilize partial checks by forming the sums of the columns of the matrices B_i; here the following relation is valid:

(8) $\Sigma_{i+1} = A\Sigma_i - p_{i+1} \begin{pmatrix} 1 \\ 1 \\ \vdots \\ 1 \end{pmatrix}; \quad \Sigma_1 = \Sigma_0 - p_1 \begin{pmatrix} 1 \\ 1 \\ \vdots \\ 1 \end{pmatrix},$

where $\sum_i$ is the column of the sums of the matrix B_i, and $\sum_0$ the analogous column of the matrix A. Finally, from the equation

$$\frac{1}{p_n} AB_{n-1} = \frac{1}{p_n} A_n = I$$

it follows that

$$(9) \qquad\qquad A^{-1} = \frac{1}{p_n} B_{n-1}.$$

Formula (9) provides an algorithm for inverting matrices. For matrices of not very high order, in case it is necessary to solve the problem of finding the proper numbers and that of inverting the matrix as well, this method is very convenient.

The number of operations necessary for obtaining the coefficients p_i (including the computation of the matrix B_n) is equal to $(n-1)n^3$ multiplications.

In Table V is shown the scheme of computation, by D. K. Faddeev's method, of the coefficients of the characteristic polynomial and the elements of the inverse matrix.

Let us pass now to the determination of the proper vectors of the matrix A.

Let the proper numbers have been computed already, and moreover let them have proved to be distinct. Let us construct the matrix

$$(10) \qquad\qquad Q_k = \lambda_k^{n-1}I + \lambda_k^{n-2}B_1 + \cdots + B_{n-1},$$

where B_i are the matrices that have been computed in the process of finding the coefficients of the characteristic polynomial, and λ_k is the kth proper number of the matrix A.

It can be proved that Q_k is a non-zero matrix, on the assumption that $\lambda_1, \ldots, \lambda_n$ are all distinct.

We shall show that each column of the matrix Q_k consists of the components of the proper vector belonging to the proper number λ_k for, indeed,

$$\begin{aligned}
(\lambda_k I - A)Q_k &= (\lambda_k I - A)(\lambda_k^{n-1}I + \lambda_k^{n-1}B_1 + \cdots + B_{n-1}) \\
&= \lambda_k^n I + \lambda_k^{n-1}(B_1 - A) + \lambda_k^{n-2}(B_2 - AB_1) + \cdots + AB_{n-1} \\
&= \lambda_k^n I - p_1\lambda_k^{n-1}I - p_2\lambda_k^{n-2}I - \cdots - q_n I = 0.
\end{aligned}$$

Hence it follows that $(\lambda_k I - A)u = 0$, where u is any column of the constructed matrix Q_k, i.e., that

(11) $$\lambda_k u = Au ,$$

which equation shows that u is the proper vector.

Observation 1. In computing the proper vectors in the manner described, there is of course no need to find all columns of the matrix Q_k. One should limit oneself to the computation of one column; its elements are obtained in the form of a linear combination of the columns of the matrices B_i bearing the same designation, with the coefficients previously given.

Observation 2. For computing the column u of the matrix Q_k the recurrence formula:

(12) $$u_0 = e; \quad u_i = \lambda_k u_{i-1} + b_i,$$

can be used conveniently; here b_i is the adopted column of the matrix B_i, and e is its counterpart column of the unit matrix.

Then

$$u = u_{n-1}.$$

As an example let us compute, for Leverrier's matrix, the proper vector that belongs to the proper number $\lambda_4 = -5.29870$.

TABLE VI.   *Determination of a Proper Vector by the Method of D. K. Faddeev*

I	II	III	IV	V	VI
2258.1433	−2990.2530	1189.8295	−148.7675	308.9523	1
70.5604	−48.1119	8.0822	0	30.5307	0.098820
32.0618	−14.2294	1.3785	0	19.2109	0.062181
4.4283	−1.5939	0.1751	0	3.0095	0.009741

In columns I, II, III are arrayed the components of the first column of the matrices B_i, multiplied by the corresponding powers of λ_4; in column IV are the components of the vector $\lambda^3(1, 0, 0, 0)$. Column V contains the components of the vector X_4; column VI, those of this vector after normalization.

§ 26. THE ESCALATOR METHOD[1]

An original method for determining the proper numbers and proper vectors of a matrix has recently appeared under the name of the *escalator method*. This method gives an inductive construction by means of which, knowing the proper numbers and proper vectors of the matrix A_{k-1} and its transpose, one can form an equation for the determination of the proper numbers of the matrix A_k— obtained from A_{k-1} by bordering—and next compute, by simple formulas, the components of the proper vectors for the matrix A_k and its transpose. The application of the escalator method is commenced by finding the proper vectors of a matrix of the second order. This problem is quite simply soluble.

The great merit of the method is the presence of a powerful check, which makes it possible for the computor to be confident at each step, not only of his computations but also of the absence of a loss of significant figures.

The method is based on a utilization of the properties of orthogonality of the proper vectors of a matrix and its transpose.

The escalator method is especially convenient in case the complete spectrum of the matrix must be found.

We shall not present the general induction from the kth step to the $(k+1)$th, but shall content ourselves with examining the transition from a third-order matrix to a fourth-order one. For convenience we shall designate the components of the vectors by different letters, contrary to established usage. We shall assume that all the proper numbers of the matrix A_3 are real and distinct.

Thus let λ_r $(r = 1, 2, 3)$ be the proper numbers of the matrices A_3 and A_3', where

$$(1) \qquad A_3 = \begin{pmatrix} a_{11} & a_{12} & a_{13} \\ a_{21} & a_{22} & a_{23} \\ a_{31} & a_{32} & a_{33} \end{pmatrix}.$$

Furthermore, let $X_r = (x_r, y_r, z_r)$ and $X_r' = (x_r', y_r', z_r')$ $(r = 1, 2, 3)$ be the aggregate of the proper vectors of these matrices.

[1] J. Morris and J. W. Head [1]; J. Morris [2].

These proper vectors may be normalized so that

(2)

$$
\begin{pmatrix}
x_1' & x_2' & x_3' \\
y_1' & y_2' & y_3' \\
z_1' & z_2' & x_3'
\end{pmatrix}
\begin{pmatrix}
x_1 & y_1 & z_1 \\
x_2 & y_2 & z_2 \\
x_3 & y_3 & z_3
\end{pmatrix}
$$

$$
=
\begin{pmatrix}
x_1 & y_1 & z_1 \\
x_2 & y_2 & z_2 \\
x_3 & y_3 & z_3
\end{pmatrix}
\begin{pmatrix}
x_1' & x_2' & x_3' \\
y_1' & y_2' & y_3' \\
z_1' & z_2' & z_3'
\end{pmatrix}
= I.
$$

This follows from the properties of orthogonality of the proper vectors of a matrix and of its transpose, established in § 3, Paragraph 12.

Let A_4 be the matrix of the fourth order obtained from A_3 by bordering; let $X = (x, y, z, u)$ be its proper vector which belongs to the proper number λ.

We have

(3)

$$
\begin{aligned}
\lambda x &= a_{11}x + a_{12}y + a_{13}z + a_{14}u \\
\lambda y &= a_{21}x + a_{22}y + a_{23}z + a_{24}u \\
\lambda z &= a_{31}x + a_{32}y + a_{33}z + a_{34}u \\
\lambda u &= a_{41}x + a_{42}y + a_{43}z + a_{44}u.
\end{aligned}
$$

Let us multiply the first three equations of system (3) by x_r', y_r', z_r', respectively, and add. We obtain:

$$
\begin{aligned}
\lambda(xx_r' + yy_r' + zz_r') &= (a_{11}x_r' + a_{21}y_r' + a_{31}z_r')x + (a_{12}x_r' + a_{22}y_r' + a_{32}z_r')y \\
&\quad + (a_{13}x_r' + a_{23}y_r' + a_{33}z_r')z + (a_{14}x_r' + a_{24}y_r' + a_{34}z_r')u,
\end{aligned}
$$

whence, on the strength of the fact that (x_r', y_r', z_r') is the proper vector for the matrix A_3':

$$
\lambda(xx_r' + yy_r' + zz_r') = \lambda_r(xx_r' + yy_r' + zz_r') + (a_{14}x_r' + a_{24}y_r' + a_{34}z_r')u
$$

and consequently

(4)

$$
xx_r' + yy_r' + zz_r' = -\frac{P_r'u}{\lambda_r - \lambda},
$$

where

(5)

$$
P_r' = a_{14}x_r' + a_{24}y_r' + a_{34}z_r'.
$$

Let

$$(6) \qquad P_r = a_{41}x_r + a_{42}y_r + a_{43}z_r.$$

Then, on the strength of the orthogonality properties (2), the following relation is valid:

$$(7) \qquad \sum_{r=1}^{3} P_r(x'_r x + y'_r y + z'_r z) = P,$$

where

$$(8) \qquad P = a_{41}x + a_{42}y + a_{43}z = -(a_{44} - \lambda)u.$$

Indeed,

$$\sum_{r=1}^{3} (a_{41}x_r + a_{42}y_r + a_{43}z_r)(x'_r x + y'_r y + z'_r z)$$
$$= a_{41}(x_1 x'_1 + x_2 x'_2 + x_3 x'_3)x + a_{41}(x_1 y'_1 + x_2 y'_2 + x_3 y'_3)y$$
$$+ a_{41}(x_1 z'_1 + x_2 z'_2 + x_3 z'_3)z + \cdots = a_{41}x + a_{42}y + a_{43}z.$$

Replacing the expression $x'_r x + y'_r y + z'_r z$ in equation (7) by $-\dfrac{P'_r u}{\lambda_r - \lambda}$ in accordance with (4), we obtain the following equation for the determination of the proper numbers of the matrix A_4:

$$(9) \qquad a_{44} - \lambda = \sum_{r=1}^{3} \frac{P_r P'_r}{\lambda_r - \lambda}.$$

Equation (9) we may well call the escalator form of the characteristic equation, or the escalator equation. Moreover, on multiplying (4) by x_r, y_r, z_r $(r = 1, 2, 3)$ successively and adding, we obtain, by once again taking into consideration the properties of orthogonality (2):

$$(10) \qquad \frac{x}{u} = -\sum_{r=1}^{3} \frac{P'_r x_r}{\lambda_r - \lambda}; \qquad \frac{y}{u} = -\sum_{r=1}^{3} \frac{P'_r y_r}{\lambda_r - \lambda};$$
$$\frac{z}{u} = -\sum_{r=1}^{3} \frac{P'_r z_r}{\lambda_r - \lambda}.$$

Analogously

$$(11) \qquad \frac{x'}{u'} = -\sum_{r=1}^{3} \frac{P_r x'_r}{\lambda_r - \lambda}; \qquad \frac{y'}{u'} = -\sum_{r=1}^{3} \frac{P_r x'_r}{\lambda_r - \lambda};$$
$$\frac{z'}{u'} = -\sum_{r=1}^{3} \frac{P_r z'_r}{\lambda_r - \lambda}.$$

Thus on finding the proper number λ from equation (9), we find, by formulas (10) and (11), the proper vectors of the matrices A_4 and A_4' belonging to this number, accurate to a constant factor. If we are to have the possibility of continuing the process, we must still normalize them in the sense of formula (2).

It may be readily verified (again using properties (2)) that

$$\frac{xx' + yy' + zz'}{uu'} = \sum_{r=1}^{3} \frac{P_r P_r'}{(\lambda_r - \lambda)^2}.$$

Consequently

$$\frac{xx' + yy' + zz' + uu'}{uu'} = 1 + \sum_{r=1}^{3} \frac{P_r P_r'}{(\lambda_r - \lambda)^2}.$$

Thus we will satisfy the normality condition with

$$\frac{1}{uu'} = 1 + \sum_{r=1}^{3} \frac{P_r P_r'}{(\lambda_r - \lambda)^2}.$$

We note that if the escalator form of the characteristic polynomial be denoted by $f(\lambda)$:

(12) $$f(\lambda) = -a_{44} + \lambda + \sum_{r=1}^{3} \frac{P_r P_r'}{\lambda_r - \lambda},$$

then

$$f'(\lambda) = 1 + \sum_{r=1}^{3} \frac{P_r P_r'}{(\lambda_r - \lambda)^2}.$$

Thus

$$\frac{1}{uu'} = f'(\lambda).$$

Without sacrificing the generality, we may consider that $u = \pm u'$, choosing the sign so that $\frac{1}{u^2} = \pm f'(\lambda)$ will be positive. This gives

(13)
$$u = u' = \frac{1}{\sqrt{f'(\lambda)}}, \quad \text{if } f'(\lambda) > 0;$$

$$u = -u' = \frac{1}{\sqrt{-f'(\lambda)}}, \quad \text{if } f'(\lambda) < 0.$$

In concluding we note the check equations

$$(14) \quad
\begin{aligned}
&\sum_{r=1}^{3} P_r x_r' = a_{41}; \quad \sum_{r=1}^{3} P_r y_r' = a_{42}; \quad \sum_{r=1}^{3} P_r z_r' = a_{43}; \\
&\sum_{r=1}^{3} P_r' x_r = a_{14}; \quad \sum_{r=1}^{3} P_r' y_r = a_{24}; \quad \sum_{r=1}^{3} P_r' z_r = a_{34}; \\
&\lambda_1 + \lambda_2 + \lambda_3 + \lambda_4 = \operatorname{tr} A_4.
\end{aligned}$$

The last equations show that all the "new" elements of the matrix A_4 are used by us for the check. A good coincidence of the check formulas guarantees the correctness of the computations at each step. Upon conclusion of the process it is useful to verify the fulfilment of the orthogonality condition by the vectors that have been found.

We have described the process only for matrices of the fourth order; the transition to the general case is obvious. Given the satisfaction of the check equations the method guarantees very great accuracy, not only for all the proper numbers, but also for the components of the proper vectors belonging to them.

In the case of a symmetric matrix, the escalator process is naturally facilitated, since all quantities marked by a prime (i.e., relating to the transposed matrix) will coincide with the corresponding quantities without primes. From the form of the escalator equation one can conclude that in this case the proper roots of the successively bordered matrices separate. This circumstance greatly facilitates the determination of the roots, which are usually found by Newton's method.

We observe that the escalator form of the characteristic equation proves to be more convenient to the application of Newton's method than the expanded form, since the computation of $f(\lambda)$ and $f'(\lambda)$ is effected very easily.

Without going into detail, we note that in case successive escalator equations have identical or complex roots, the process described above must be somewhat modified.[1]

We shall find the proper numbers and proper vectors of the Leverrier matrix by the escalator method.

[1] J. Morris and J. W. Head [1].

The solution will consist of three stages.

Stage I.   For the matrix A_2,

$$\begin{pmatrix} -5.509882 & 1.870086 \\ 0.287865 & -11.811654 \end{pmatrix},$$

the equation for the determination of the proper numbers will be

$$\lambda^2 + 17.321536\lambda + 64.542487 = 0;$$

its roots:

$$\lambda_1 = -11.895952, \quad \lambda_2 = -5.425584.$$

For the check we form $\lambda_1 + \lambda_2 = -17.321536$, and compute the trace of the matrix A_2:

$$\text{tr } A_2 = -17.321536.$$

Following this we compute the proper vectors of the matrices A_2 and A_2', solving the corresponding systems and normalizing the vectors obtained:

X_1	X_2	X_1'	X_2'	
-0.061767	4.679234	-0.210926	0.210926	
0.210926	0.210926	4.679234	0.061767	I
0.905643	1.138422	26.638114	0.442009	P_i and P_i'

The first stage has been completed.

Stage II.   We form the matrix

$$\begin{pmatrix} -5.509882 & 1.870086 & 0.422908 \\ 0.287865 & -11.811654 & 5.711900 \\ 0.049099 & 4.308033 & -12.970687 \end{pmatrix}.$$

We copy on a separate sheet the newly introduced coefficients a_{13}, a_{23} and a_{31}, a_{32} in the form of columns, and juxtaposing them to the columns of the proper vectors of the matrix A_r, we find the quantities P_i' and P_i (by accumulation). For convenience in future computations we copy them along with the proper vectors in scheme (I), arranging them in a row.

We can now write the escalator equation for the matrix A_3:

$$f(\lambda) = 12.970687 + \lambda + \frac{24.124621}{-11.895952 - \lambda} + \frac{0.503193}{-5.425584 - \lambda} = 0.$$

Let us determine its roots by Newton's method, arranging the computation in accordance with the scheme:

λ	-15	-16.651	-17.3458	-17.3975	-17.397655	
$-11.895952 - \lambda$	3.104	4.755	5.4498	5.501548	5.501703	
$-5.425584 - \lambda$	9.574	11.225	11.9202	11.971916	11.972071	
$12.970687 + \lambda$	-2.029	-3.680	-4.3751	-4.426813	-4.426968	
$\dfrac{P_1 P_1'}{-11.895958 - \lambda}$	7.772	5.074	4.4267	4.385061	4.384936	
$\dfrac{P_2 P_2'}{-5.425584 - \lambda}$	0.053	0.045	0.0422	0.042031	0.042031	
$f(\lambda)$	5.796	1.439	0.0938	0.000279	-0.000001	II
$\dfrac{P_1 P_1'}{(-11.895952 - \lambda)^2}$	2.504	1.067	0.8123	0.797059	0.797014	
$\dfrac{P_2 P_2'}{(-5.425584 - \lambda)^2}$	0.006	0.004	0.0035	0.003511	0.003511	
$f'(\lambda)$	3.510	2.071	1.8158	1.800570	1.800525	
$\Delta\lambda$	-1.651	-0.6948	-0.0517	-0.000155	0.000000	

Thus $\qquad\qquad\qquad\qquad \lambda_1 = -17.397655.$

Analogously we find $\lambda_2 = -7.594378$ and $\lambda_3 = -5.300190$. The check: $\lambda_1 + \lambda_2 + \lambda_3 = -30.292223.$

$$\mathrm{tr}\, A_3 = -30.292223.$$

We now pass on to the determination of the components of the proper vectors of the matrices A_3 and A_3', which are found, accurate to a constant factor, by formulas analogous to formulas (10) and (11).

While so doing it is handy to set up auxiliary schemes (III) and (IV).

$P'_i x_i$	$P'_i y_i$	$P_i x'_i$	$P_i y'_i$	
-1.645356	5.618671	-0.191024	4.237716	III
2.068264	0.093231	0.240123	0.070317	
0.422908	5.711902	0.049099	4.308033	

$\dfrac{1}{\lambda_i + 17.397655}$	$\dfrac{1}{\lambda_i + 7.594378}$	$\dfrac{1}{\lambda_i + 5.300190}$	
0.181762	-0.232473	-0.151613	IV
0.083528	0.461086	-7.974863	

Scheme (III) contains the corresponding products of the numbers P and the components of the vectors, and is obtained from Scheme (I); in the last row the check is carried out (for example $\sum\limits_{i=1}^{2} P'_i x_i = a_{13}$); Scheme (IV) contains the factor $\dfrac{1}{\lambda_i - \lambda}$, where for λ one takes the three computed roots in succession.

The normalizing factors are now determined from Scheme (II) and the two analogous schemes serving for the computation of the two other roots. Since $f'(\lambda_i) > 0$,

$$D_i = \frac{1}{\sqrt{f'(\lambda_i)}} \text{ and } z_i = z'_i = D_i; \quad i = 1, 2, 3.$$

On computing these, we obtain $D_1 = 0.745248$, $D_2 = 0.644055$, $D_3 = 0.172627$.

Utilizing the preceding schemes, we easily find the components of the proper vectors of the matrices A_3 and A'_3 (in the final scheme we copy them after multiplication by the corresponding normalizing factor).

X_1	X_2	X_3	X_1'	X_2'	X_3'	
0.094129	−0.860553	2.804268	0.010928	−0.099909	0.325572	
−0.766896	0.813572	0.275404	−0.578409	0.613612	0.207717	
0.745248	0.644055	0.172627	0.745248	0.644055	0.172627	
0.835026	1.114160	0.333026	0.137039	0.182847	0.054654	P_i and P_i'

The second stage has been completed.

Stage III. Having computed the quantities P_i and P_i', we write the escalator equation for the matrix A_4:

$$17.596207 + \lambda + \frac{0.114431}{-17.397655 - \lambda} + \frac{0.203721}{-7.594378 - \lambda}$$
$$+ \frac{0.018201}{-5.300190 - \lambda} = 0$$

and compute its roots:

$$\lambda_1 = -17.863262, \qquad \lambda_2 = -17.152427,$$
$$\lambda_3 = -7.574044, \qquad \lambda_4 = -5.298698,$$
$$\sum_{i=1}^{4} \lambda_i = -47.888431; \quad \text{tr } A_4 = -47.888430.$$

We next compute the proper vectors of the matrix A_4, normalizing them as usual:

X_1	X_2	X_3	X_4
−0.019872	0.032932	−0.351235	1.135218
0.169807	−0.261310	0.328467	0.112183
−0.187215	0.236640	0.260927	0.070591
0.808482	0.586694	0.045005	0.011058

and the proper vectors of the matrix A_4':

X_1'	X_2'	X_3'	X_4'
−0.014058	0.023297	−0.248476	0.803091
0.780383	−1.200908	1.509559	0.515564
−1.140762	1.441927	1.589924	0.430123
0.808482	0.586694	0.045005	0.011058

The final check of the computation is the computation of the product of the two matrices composed respectively of the components of the proper vectors. Instead of the unit matrix there is obtained the matrix

$$\begin{pmatrix} 1.000005 & -0.000004 & 0.000000 & 0.000002 \\ -0.000003 & 1.000004 & -0.000002 & -0.000003 \\ -0.000002 & 0.000000 & 0.999994 & 0.000000 \\ -0.000001 & 0.000000 & 0.000004 & 1.000006 \end{pmatrix}.$$

Lastly, for comparison, we state the escalator equation in the ordinary polynomial form:

$$\lambda^4 + 47.888430\lambda^3 + 797.27877\lambda^2 + 5349.4556\lambda + 12296.550 = 0,$$

and we normalize the proper vector belonging to λ_4 so that its first component is equal to unity. This gives X_4 (1; 0.098820; 0.062183; 0.009741).

We see that the coefficients of the equation here cited coincide with Leverrier's data with greater accuracy than do the computations by the other methods. The orthogonality relation is also satisfied with considerable accuracy.

§ 27. THE METHOD OF INTERPOLATION

The methods that we have presented in the preceding sections solved the problem of bringing the secular equation into polynomial form. The interpolation method, developed in this section, is applicable to a more general case, to wit, the expansion of a determinant of the form

$$(1) \qquad F(\lambda) = \begin{vmatrix} f_{11}(\lambda) & \cdots & f_{1n}(\lambda) \\ \cdot & \cdots & \cdot \\ \cdot & \cdots & \cdot \\ f_{n1}(\lambda) & \cdots & f_{nn}(\lambda) \end{vmatrix},$$

($f_{ik}(\lambda)$ is a given polynomial in λ), in particular, to the expansion of the characteristic determinant $D(\lambda) = |A - \lambda I|$ and of the determinant $|A - B\lambda|$ where A and B are given matrices.

The essence of the method consists in the following. Let it be known that $F(\lambda)$ is a polynomial of degree not exceeding the number k. As is known from higher algebra, such a polynomial is completely determined by its values at $k+1$ points and may be reconstructed from such values by means of one or another interpolation formula.

For an explicit representation of $F(\lambda)$, therefore, it is necessary to compute the value of $k+1$ numerical determinants

$$(2) \qquad F(\lambda_i) = \begin{vmatrix} f_{11}(\lambda_i) & \cdots & f_{1n}(\lambda_i) \\ \cdot & \cdots & \cdot \\ f_{n1}(\lambda_i) & \cdots & f_{nn}(\lambda_i) \end{vmatrix}, \quad i = 0, 1, \ldots, k$$

where $\lambda_0, \lambda_1, \ldots, \lambda_k$ are certain numbers chosen arbitrarily, generally speaking.

The computation of the necessary determinants can be accomplished, for instance, by the scheme expounded in § 7.

For the construction of the polynomial $F(\lambda)$ by its values it is most convenient to use the Newton interpolation formula applicable for equally spaced abscissas λ_i.

We cite the Newton formula for $\lambda_i = i$, $i = 0, \ldots, k$:

$$(3) \qquad F(\lambda) = \sum_{i=0}^{k} \frac{\Delta^i F(0)}{i!} \lambda(\lambda-1) \ldots (\lambda-i+1),$$

where $\Delta^i F(l)$ designates the ith difference of the computed values of the polynomial $F(\lambda)$, which is determined by the recurrence formula

$$\Delta^i F(l) = \Delta^{i-1} F(l+1) - \Delta^{i-1} F(l).$$

Let us put

$$\frac{\lambda(\lambda-1)\ldots(\lambda-i+1)}{i!} = \sum_{m=1}^{i} c_{mi} \lambda^m.$$

Then formula (3) is transformed into the form

(4)
$$F(\lambda) = \sum_{i=0}^{k} \Delta^i F(0) \left(\sum_{m=1}^{i} c_{mi} \Delta^i F(0) \right) \lambda^m$$

$$= F(0) + \sum_{m=1}^{k} \left(\sum_{i=m}^{k} c_{mi} \Delta^i F(0) \right) \lambda^m.$$

This formula bears the appellation "the A. A. Markov interpolation formula".

In Sh. E. Mikeladze's work [1] formula (4) has been chosen as the interpolation formula.

We attach a table of the coefficients c_{mi} for $m \leqslant i \leqslant 20$, expressed as decimal fractions.[1]

In using the interpolation formula it is convenient as a check of the supporting values of determinant (1) to compute one more value of $F(\lambda)$, viz., in our case, $F(k+1)$, for $\Delta^{k+1} F(0)$ must be equal to zero, and $\Delta^k F(0)$ and $\Delta^k F(1)$ are equal to each other.

The interpolation method requires a great many operations. Thus for computing the coefficients of the characteristic polynomial by means of interpolation formula (4), it is first of all necessary to compute $(n+1)$ determinants of the nth order. This requires $\frac{n+1}{3}(n-1)(n^2+n+3)$ multiplications and divisions. If one takes the coefficients of the interpolation formula from a table, it is still necessary to carry out $\frac{n(n+1)}{2}$ multiplications to obtain the coefficients of the characteristic polynomial.

[1] This table was computed by N. M. Terentiev and K. I. Grishmanovskaya.

The overall number of multiplication and division operations is thus

$$\frac{n+1}{3}(n-1)(n^2+n+3)+\frac{n(n+1)}{2}$$

which exceeds by far the number of operations necessary for computing the same coefficients by the method of A. M. Danilevsky or by that of A. N. Krylov.

In addition, the method in question does not permit one to simplify in any way the problem of finding the proper vectors of the matrix, whereas in computing by the methods of A. M. Danilevsky or A. N. Krylov the task of determining the proper vectors of the matrix is much facilitated. Nevertheless the interpolation method is interesting as a method making possible the solution of more general problems.

As an example we again exhibit a computation for Leverrier's matrix.

Omitting the wearisome computation of the determinants, we find that

$$D(0) = 12296.55, \quad D(1) = 18492.17, \quad D(2) = 26583.68,$$
$$D(3) = 36894.41, \quad D(4) = 49771.69.$$

We next compose a table of differences:

λ	$D(\lambda)$	Δ	Δ^2	Δ^3	Δ^4
0	12296.55				
		6195.62			
1	18492.17		1895.89		
		8091.51		323.33	
2	26583.68		2219.82		24
		10310.73		347.33	
3	36894.41		2566.55		
		12877.28			
4	49771.69				

(We note that in case one is computing the coefficients of a characteristic polynomial, the computation of superfluous values of $D(\lambda)$ for a check need not be done, since in this case a reliable check is the equality $\Delta^k F(0) = (-1)^n n!$).

TABLE VII. Table of Coefficients of A. A. Markov's Interpolation Formula

i \ m	1	2	3	4	5	6
1	1.00000 00000					
2	−0.50000 00000	0.50000 00000				
3	0.33333 33333	−0.50000 00000	0.16666 66667			
4	−0.25000 00000	0.45833 33333	−0.25000 00000	0.041666 66667		
5	0.20000 00000	−0.41666 66667	0.29166 66667	−0.083333 33333	$0.83333\ 33333 \cdot 10^{-2}$	
6	−0.16666 66667	0.38055 55556	−0.31250 00000	0.11805 55556	$-0.20833\ 33333 \cdot 10^{-1}$	$0.13888\ 88889 \cdot 10^{-2}$
7	0.14285 71429	−0.35000 00000	0.32222 22222	−0.14583 33333	$0.34722\ 22222 \cdot 10^{-1}$	$-0.41666\ 66667 \cdot 10^{-2}$
8	−0.12500 00000	0.32410 71429	−0.32569 44444	0.16788 19444	$-0.48611\ 11111 \cdot 10^{-1}$	$0.79861\ 11111 \cdot 10^{-1}$
9	0.11111 11111	−0.30198 41270	0.32551 80776	−0.18541 66667	$0.61863\ 42593 \cdot 10^{-1}$	$-0.12500\ 00000 \cdot 10^{-1}$
10	−0.10000 00000	0.28289 68254	−0.32316 46825	0.19942 68078	$-0.74218\ 75000 \cdot 10^{-1}$	$0.17436\ 34259 \cdot 10^{-1}$
11	0.090909 09091	−0.26626 98413	0.31950 39683	−0.21067 57055	$0.85601\ 30071 \cdot 10^{-1}$	$-0.22598\ 37963 \cdot 10^{-1}$
12	−0.083333 33333	0.25165 64454	−0.31506 77910	0.21974 47274	$-0.96024\ 16777 \cdot 10^{-1}$	$0.27848\ 62305 \cdot 10^{-1}$
13	0.076923 07692	−0.23870 85137	0.31018 99952	−0.22707 72707	0.10554 11339	$-0.33092\ 89572 \cdot 10^{-1}$
14	−0.071428 57143	0.22715 24111	−0.30508 41751	0.23301 38939	−0.11422 22865	$0.38267\ 76988 \cdot 10^{-1}$
15	0.066666 66667	−0.21677 08218	0.29988 87242	−0.23781 85793	0.12214 17270	$-0.43331\ 40432 \cdot 10^{-1}$
16	−0.062500 00000	0.20738 93121	−0.29469 38553	0.24169 79634	−0.12937 15303	$0.48257\ 04949 \cdot 10^{-1}$
17	0.058823 52941	−0.19886 64114	0.28955 82939	−0.24481 53688	0.13597 89675	$-0.53028\ 48954 \cdot 10^{-1}$
18	−0.055555 55556	0.19108 62513	−0.28451 98560	0.24730 10868	−0.14202 54343	$0.57636\ 84942 \cdot 10^{-1}$
19	0.052631 57895	−0.18395 30567	0.27960 22978	−0.24925 99694	0.14756 62581	$-0.62078\ 35389 \cdot 10^{-1}$
20	−0.050000 00000	0.17738 69829	−0.27481 98358	0.25077 70858	−0.15265 09436	$0.66352\ 74910 \cdot 10^{-1}$

TABLE VII. *Table of Coefficients of A. A. Markov's Interpolation Formula*—Continued

m / i	7	8	9	10	11
1					
2					
3					
4					
5					
6					
7	$0.19841\ 26984 \cdot 10^{-3}$				
8	$-0.69444\ 44444 \cdot 10^{-3}$	$0.24801\ 58730 \cdot 10^{-4}$			
9	$0.15046\ 29630 \cdot 10^{-2}$	$-0.99206\ 34921 \cdot 10^{-4}$	$0.27557\ 31922 \cdot 10^{-5}$		
10	$-0.26041\ 66667 \cdot 10^{-2}$	$0.23974\ 86772 \cdot 10^{-3}$	$-0.12400\ 79365 \cdot 10^{-4}$	$0.27557\ 31922 \cdot 10^{-6}$	
11	$0.39525\ 46296 \cdot 10^{-2}$	$-0.45469\ 57672 \cdot 10^{-3}$	$0.33068\ 78307 \cdot 10^{-4}$	$-0.13778\ 65961 \cdot 10^{-5}$	$0.25052\ 10839 \cdot 10^{-7}$
12	$-0.55063\ 65741 \cdot 10^{-2}$	$0.74618\ 33113 \cdot 10^{-3}$	$-0.68204\ 36508 \cdot 10^{-4}$	$0.40187\ 75720 \cdot 10^{-5}$	$-0.13778\ 65961 \cdot 10^{-6}$
13	$0.72250\ 00919 \cdot 10^{-2}$	$-0.11123\ 51190 \cdot 10^{-2}$	$0.12035\ 65917 \cdot 10^{-3}$	$-0.89561\ 28748 \cdot 10^{-5}$	$0.43632\ 42210 \cdot 10^{-6}$
14	$-0.90727\ 07690 \cdot 10^{-2}$	$0.15489\ 69028 \cdot 10^{-2}$	$-0.19121\ 33488 \cdot 10^{-3}$	$0.16913\ 30467 \cdot 10^{-4}$	$-0.10448\ 81687 \cdot 10^{-5}$
15	$0.11019\ 04517 \cdot 10^{-1}$	$-0.20505\ 51606 \cdot 10^{-2}$	$0.28173\ 03941 \cdot 10^{-3}$	$-0.28533\ 30761 \cdot 10^{-4}$	$0.21027\ 76553 \cdot 10^{-5}$
16	$-0.13038\ 56762 \cdot 10^{-1}$	$0.26110\ 82453 \cdot 10^{-2}$	$-0.39228\ 17198 \cdot 10^{-3}$	$0.44358\ 12552 \cdot 10^{-4}$	$-0.37546\ 84744 \cdot 10^{-5}$
17	$0.15110\ 24302 \cdot 10^{-1}$	$-0.32244\ 63934 \cdot 10^{-2}$	$0.52279\ 94100 \cdot 10^{-3}$	$-0.64824\ 21930 \cdot 10^{-4}$	$0.61431\ 22437 \cdot 10^{-5}$
18	$-0.17216\ 81227 \cdot 10^{-1}$	$0.38847\ 84994 \cdot 10^{-2}$	$-0.67289\ 18835 \cdot 10^{-3}$	$0.90267\ 28545 \cdot 10^{-4}$	$-0.94031\ 83373 \cdot 10^{-5}$
19	$0.19344\ 18265 \cdot 10^{-1}$	$-0.45864\ 70640 \cdot 10^{-2}$	$0.84193\ 88893 \cdot 10^{-3}$	$-0.12093\ 17380 \cdot 10^{-3}$	$0.13659\ 18875 \cdot 10^{-4}$
20	$-0.21480\ 89121 \cdot 10^{-1}$	$0.53243\ 56241 \cdot 10^{-2}$	$-0.10291\ 65477 \cdot 10^{-2}$	$0.15698\ 20955 \cdot 10^{-3}$	$-0.19022\ 81621 \cdot 10^{-4}$

TABLE VII. *Table of Coefficients of A. A. Markov's Interpolation Formula*—Continued

i \ m	12	13	14	15	16
1					
2					
3					
4					
5					
6					
7					
8					
9					
10					
11					
12	$0.20876\ 75699 \cdot 10^{-8}$				
13	$-0.12526\ 05419 \cdot 10^{-7}$	$0.16059\ 04384 \cdot 10^{-9}$			
14	$0.42797\ 35183 \cdot 10^{-7}$	$-0.10438\ 37849 \cdot 10^{-8}$	$0.11470\ 74560 \cdot 10^{-10}$		
15	$-0.10960\ 29742 \cdot 10^{-6}$	$0.38274\ 05448 \cdot 10^{-8}$	$-0.80295\ 21918 \cdot 10^{-10}$	$0.76471\ 63732 \cdot 10^{-12}$	
16	$0.23417\ 63229 \cdot 10^{-6}$	$-0.10438\ 37849 \cdot 10^{-7}$	$0.31448\ 96085 \cdot 10^{-9}$	$-0.57353\ 72799 \cdot 10^{-11}$	$0.47794\ 77332 \cdot 10^{-13}$
17	$-0.44126\ 50535 \cdot 10^{-6}$	$0.23599\ 43405 \cdot 10^{-7}$	$-0.91001\ 24841 \cdot 10^{-9}$	$0.23897\ 38666 \cdot 10^{-10}$	$-0.38235\ 81866 \cdot 10^{-12}$
18	$0.75803\ 49082 \cdot 10^{-6}$	$-0.46803\ 07957 \cdot 10^{-7}$	$0.21705\ 35904 \cdot 10^{-8}$	$-0.73126\ 00319 \cdot 10^{-10}$	$0.16887\ 48657 \cdot 10^{-11}$
19	$-0.12130\ 42722 \cdot 10^{-5}$	$0.84236\ 33371 \cdot 10^{-7}$	$-0.45196\ 17150 \cdot 10^{-8}$	$0.18351\ 59980 \cdot 10^{-9}$	$-0.54486\ 04159 \cdot 10^{-11}$
20	$0.18353\ 50023 \cdot 10^{-5}$	$-0.14067\ 66531 \cdot 10^{-6}$	$0.85054\ 52978 \cdot 10^{-8}$	$-0.40032\ 10556 \cdot 10^{-9}$	$0.14351\ 97385 \cdot 10^{-10}$

TABLE VII. *Table of Coefficients of A. A. Markov's Interpolation Formula*—Continued

m / i	17	18	19	20
1				
2				
3				
4				
5				
6				
7				
8				
9				
10				
11				
12				
13				
14				
15				
16				
17	$0.28114\ 57254 \cdot 10^{-14}$			
18	$-0.23897\ 38666 \cdot 10^{-13}$	$0.15619\ 20697 \cdot 10^{-15}$		
19	$0.11152\ 11378 \cdot 10^{-12}$	$-0.14057\ 28627 \cdot 10^{-14}$	$0.82206\ 35247 \cdot 10^{-17}$	
20	$-0.37837\ 52888 \cdot 10^{-12}$	$0.69114\ 99084 \cdot 10^{-14}$	$-0.78096\ 03484 \cdot 10^{-16}$	$0.41103\ 17623 \cdot 10^{-18}$

Lastly we compute the coefficients of the characteristic polynomial, arranging the computation in accordance with the scheme:

i	$\Delta^i D(0)$	c_{4i}	c_{3i}	c_{2i}	c_{1i}
1	6195.62				1.00000000
2	1895.89			0.50000000	-0.50000000
3	323.33		0.16666667	-0.50000000	0.33333333
4	24.00	0.04166667	-0.25000000	0.45833333	-0.25000000
			47.8883	797.280	5349.45

The coefficient $p_4 = D(0) = 12296.55$.

The interpolation method is always applicable; in particular the case in which the characteristic polynomial has multiple roots is not a whit distinguished from the other cases.

If instead of the numbers $0, \ldots, k$ the numbers $\lambda_i = a + hi$ be taken as the interpolation nodes, formula (4) will thereupon be changed in form as follows:

$$(5) \qquad F(\lambda) = F(a) + \sum_{m=1}^{k} \left(\sum_{i=m}^{k} c_{mi} h^i \Delta^i F(a) \right) (\lambda - a)^m.$$

It may sometimes be convenient to take as interpolation abscissas numbers that are not equally spaced. In this case one can use the general interpolation formula of Newton's. However in case of abscissas not equally spaced it is more convenient to construct the required polynomial by the method of undetermined coefficients, viz.: =

$$F(\lambda) = a_0 \lambda^k + a_1 \lambda^{k-1} + \cdots + a_k.$$

Then for the determination of the numbers $a_j, j = 0, \ldots, k$, we will obtain a system of algebraic equations

$$F(\lambda_i) = a_0 \lambda_i^k + a_1 \lambda_i^{k-1} + \cdots + a_k,$$

which may be solved by any of the methods set forth previously.

The interpolation method may be employed conveniently, in particular, to the expansion of the determinant $|A - B\lambda|$ in case the matrix B has a small determinant. If, however, the determinant of B is not a small number, the coefficients of the desired polynomial can better be determined by means of the transformation

$$|A - B\lambda| = |B| \, |AB^{-1} - \lambda I|.$$

The matrix AB^{-1} may be found by the elimination method (§ 12).

§ 28. COMPARISON OF THE METHODS

In the foregoing sections there have been expounded six different methods for bringing the characteristic determinant of the matrix into polynomial form; the determination of the proper numbers of a matrix is thus reduced to the determination of the roots of an algebraic equation. Four of these methods also make possible the determination of the proper vectors belonging to the proper numbers of the matrix, avoiding the solution of the linear systems defining the components of these vectors.

We shall endeavor to characterize the peculiarities of these methods.

The least number of operations for the entire process of computation is required by the method of A. M. Danilevsky. The Danilevsky scheme secures a comparatively high accuracy for the sought-for coefficients, which are all determined simultaneously and therefore with approximately the same accuracy. The coincidence of the coefficient p_1 with the trace of the matrix serves as a good final check of the accuracy. The method makes possible the determination of the proper vectors easily enough, too. The somewhat complicated "pattern" of the computational scheme is easily mastered by the computor. Lastly, the presence of multiple roots of the characteristic equation is no obstacle to the use of the method.

The method of A. N. Krylov requires a somewhat greater volume of computations than does that of A. M. Danilevsky. This method can nonetheless be recommended for wide application thanks to the simplicity and compactness of the computational scheme.

There is a certain shortcoming of the method: in many cases the linear system for the determination of the unknown coefficients proves to be inconvenient owing to the coefficients of this system being of different orders of magnitude. This may lower the accuracy of the computation. Moreover in case there are multiple roots of the characteristic equation, the method permits the determination of the coefficients of the minimum polynomial only.

The Samuelson method differs little in its idea from the A. N. Krylov method. Its computational scheme may sometimes be useful thanks to the complete uniformity of operations.

The escalator method solves a somewhat more general problem than do any of the rest of the methods described, viz.: we find not only all the proper numbers of the matrix, but also all the proper vectors of the matrix itself and of its transpose as well, by the very nature of the process. The overall number of operations necessary for the solution of this problem cannot be reckoned, since in the course of the process one has to find the roots of an $(n-1)$th algebraic equation of degree from two to n. And the solution of the algebraic equations—which is best carried out by Newton's method—can require a different number of operations in different cases. The escalator method is, unconditionally, substantially more labor-consuming than are the three methods considered above, but it makes it possible to obtain significant accuracy and guarantees the reliability of the result.

The method of Leverrier, and the interpolation method, can hardly compete with the first three methods described, for the number of operations required for these two methods is substantially greater. The value of the interpolation method consists in its generality. The method of Leverrier in the modification presented here is convenient for matrices of low orders.

§ 29. DETERMINATION OF THE FIRST PROPER NUMBER OF A MATRIX. FIRST CASE

In connection with problems of mathematical physics it is often necessary to find only a few of the first proper numbers of greatest modulus, and only the first proper number need be found to a significant degree of accuracy.

To the solution of this problem it is convenient to apply iterative methods that in respect of technique recall the iterative methods for the solution of a system of linear equations. In using iterative methods the dominant proper number is obtained in the form of a limit of some sequence that is constructed by a uniform process. The iterative method also gives one the possibility of computing the proper vectors belonging to the proper numbers of largest modulus.

In expounding the iterative methods we shall not consider the question in all its generality, but shall limit ourselves to a discussion of certain cases that are particularly important in applications.

In this section we shall assume that the dominant proper number of the matrix is real, and that the elementary divisors connected with it are linear. For simplicity of exposition we shall assume in addition that all the rest of the elementary divisors of the matrix are linear, although the derivations that we shall make would hold even without this assumption.

The case we consider is that met most frequently in practical problems. In particular, symmetric matrices are subsumed under this case, as was established in § 4, as are also matrices all of whose proper numbers are distinct.

On the strength of the linearity of the elementary divisors, to each proper number of the matrix A will correspond as many linearly independent vectors as its multiplicity. Let $\lambda_1, \lambda_2, \ldots, \lambda_n$ be the proper numbers of the matrix A, arranged in order of diminishing modulus (some of them perhaps equal); let $X_1, X_2, \ldots, X_n$ be the proper vectors corresponding to them. If all the numbers λ_i are distinct, the vectors X_i are determinate but for a constant factor; if two of the proper numbers are equal, any linear combination of the proper vectors belonging to these numbers will also be a proper vector. It is therefore possible so to choose the proper vectors of a matrix and its transpose that the orthogonality and normality relations (in the sense of § 3, Paragraph 12) are satisfied.

An arbitrary vector Y_0 may in this case be represented uniquely in the form

$$(1) \qquad Y_0 = a_1 X_1 + a_2 X_2 + \cdots + a_n X_n,$$

where the numbers a_i are constants, some perhaps zero.

Form the sequence of vectors

$$(2) \qquad A Y_0, A^2 Y_0, \ldots, A^k Y_0, \ldots .$$

We shall call the vector $A^k Y_0$ the kth *iterate* of the vector Y_0 *by the matrix A.*

Obviously

$$(3) \qquad AY_0 = a_1\lambda_1 X_1 + a_2\lambda_2 X_2 + \cdots + a_n\lambda_n X_n,$$

$$\cdots \cdots \cdots \cdots \cdots \cdots \cdots \cdots \cdots$$

$$(4) \qquad A^k Y_0 = a_1\lambda_1^k X_1 + a_2\lambda_2^k X_2 + \cdots + a_n\lambda_n^k X_n.$$

We use the notation $A^k Y_0 = Y_k = (y_{1k}, y_{2k}, \ldots, y_{nk})$, and shall elucidate the structure of the components of Y_k. Let

$$X_1 = (x_{11}, x_{21}, \ldots, x_{n1}),$$

$$X_2 = (x_{12}, x_{22}, \ldots, x_{n2}),$$

$$\cdots \cdots \cdots \cdots \cdots \cdots$$

$$X_n = (x_{1n}, x_{2n}, \ldots, x_{nn}).$$

We then obtain from (4)

$$(5) \qquad y_{ik} = a_1 x_{i1} \lambda_1^k + a_2 x_{i2} \lambda_2^k + \cdots + a_n x_{in} \lambda_n^k,$$

where the coefficient of λ_1^k reduces to zero when, and only when, $a_1 = 0$ or $x_{i1} = 0$. The coefficients of λ_1^k in all the components of the vector Y_k can equal zero only if $a_1 = 0$, since the vector X_1 is not zero.

Thus any component of the vector Y_k depends linearly on $\lambda_1^k, \ldots, \lambda_n^k$. Let us denote any of these components by y_k (omitting the first index).

Then

$$(6) \qquad y_k = c_1\lambda_1^k + c_2\lambda_2^k + \cdots + c_n\lambda_n^k,$$

where the coefficients c_i do not depend on the index k.

If among the numbers λ_i some are equal, the corresponding terms may be combined. Thus if λ_1 is a proper number of multiplicity r, then

$$(7) \qquad y_k = c_1\lambda_1^k + c_{r+1}\lambda_{r+1}^k + \cdots + c_n\lambda_n^k.$$

We shall now assume that the vector Y_0 has been so chosen that $a_1 \neq 0$[1]. Then λ_1^k appears with coefficients $c_1 \neq 0$ in at least one component of the kth iterate of the vector Y_0.

---

[1] Since $a_1 = (X_1', Y_0)$, where X_1' is the first proper vector of the transposed matrix, the requirement that $a_1 \neq 0$ will be fulfilled in case the vector Y_0 is not orthogonal to the vector X_1'.

Let us consider three cases:

$$1) \quad |\lambda_1| > |\lambda_2|$$

$$2) \quad \lambda_1 = \lambda_2 = \cdots = \lambda_r; \quad |\lambda_r| > |\lambda_{r+1}|$$

$$3) \quad \lambda_1 = -\lambda_2; \quad |\lambda_1| > |\lambda_3|.$$

In the first case we have

(8)
$$\frac{y_{k+1}}{y_k} = \frac{c_1\lambda_1^{k+1} + c_2\lambda_2^{k+1} + \cdots + c_n\lambda_n^{k+1}}{c_1\lambda_1^k + c_2\lambda_2^k + \cdots + c_n\lambda_n^k}$$

$$= \lambda_1 \frac{1 + b_2\alpha_2^{k+1} + b_3\alpha_3^{k+1} + \cdots + b_n\alpha_n^{k+1}}{1 + b_2\alpha_2^k + b_3\alpha_3^k + \cdots + b_n\alpha_n^k},$$

where

(9)
$$b_i = \frac{c_i}{c_1}, \quad \alpha_i = \frac{\lambda_i}{\lambda_1}.$$

Carrying out the division and retaining the terms up to the order of α_2^{2k} and α_3^k inclusive, we obtain

(10)
$$\frac{y_{k+1}}{y_k} = \lambda_1[1 - b_2'\alpha_2^k - b_3'\alpha_3^k + b_2b_2'\alpha_2^{2k}] + O(\alpha_3^k + \alpha_2^{2k}),$$

where

(11)
$$b_2' = b_2(1-\alpha_2), \quad b_3' = b_3(1-\alpha_3).$$

Hence we see that if k is sufficiently large,

(12)
$$\lambda_1 \simeq \frac{y_{k+1}}{y_k},$$

i.e., the first proper number is approximately equal to the ratio of any corresponding components of two neighboring and sufficiently high iterates of an arbitrary vector by the matrix A.

In carrying out the iterations in practice, one should compute the ratio $\frac{y_{k+1}}{y_k}$ for several components. A good coincidence of these ratios will show that in expression (10) the difference of values of the coefficients b_2', b_3' has already ceased to play a significant role.

The rapidity of the convergence of the indicated iterative process is determined by the magnitude of the ratio $\frac{\lambda_2}{\lambda_1}$, and may be very slow.

Sometimes in computing the iterates it is expedient to divide at each step the components of the vectors being iterated by the first or largest component, or to normalize them in the customary manner, to avoid growth of the components. In doing this, rather than the sequence Y_k we obtain a sequence $\tilde{Y}_k = \mu_k A^k Y_0$, where μ_k is the normalizing factor, and to obtain λ_1 one must take the ratio of the components of the vectors $A\tilde{Y}_k$ and $\tilde{Y}_k$.

It may occur, although this is improbable, that the initial vector Y_0 has been unfortunately chosen, viz.: such that the coefficient a_1 is equal to zero, or is very close to zero. In this case, at the first steps of the iteration the preponderant term will be that dependent on λ_2 (if $a_2 \neq 0$). However even if a_1 is equal to zero exactly, after several steps of the iteration the summand dependent on λ_1, thanks to rounding errors, will appear at first with a very small coefficient; in proportion as the iterations proceed, this summand will grow with satisfactory rapidity by comparison with the rest. A "struggle for dominance" of the terms depending on λ_1 and λ_2 much darkens the picture and makes necessary the replacement of the initial vector.

Example 1. Let us try to determine the first proper number of the Leverrier matrix.

We will take the vector $(1, 0, 0, 0)$ for the initial one and form 20 iterates, normalizing them at each step.

We subjoin only the last two iterates:

$\tilde{Y}_{19}$	$A\tilde{Y}_{19}$	Ratio of the components
1.00000	-17.4655	-17.466
-8.20321	143.3809	-17.479
8.17013	-143.0881	-17.514
-7.95957	149.2676	-18.753
-6.99265	132.0949	

From the table we see that the ratios of the different components are still far from each other; this shows that the process has not yet stabilized. Indeed, accurate to three decimals, $\lambda_1 = -17.863$. The process of iteration converges slowly because of the fact that the second proper number $\lambda_2 = -17.152$ differs little from the first in modulus.

Example 2. Let us determine the first proper number of the matrix

$$\begin{pmatrix} -5.509882 & 1.870086 & 0.422908 \\ 0.287865 & -11.811654 & 5.711900 \\ 0.049099 & 4.308033 & -12.970687 \end{pmatrix}.$$

We shall take $(1, 0, 0)$ as the initial vector. We display a table of the iterates commencing with the 12th:

$\tilde{Y}_{12}$	$A\tilde{Y}_{12}$	$\tilde{Y}_{13}$	$A\tilde{Y}_{13}$	$\tilde{Y}_{14}$	$A\tilde{Y}_{14}$
1.0000000	−17.351783	1.0000000	−17.378482	1.0000000	−17.389552
−8.1139091	141.126754	−8.1332710	141.483894	−8.1413264	141.632991
7.8783245	−137.093170	7.9008117	−137.468256	7.9102568	−137.625469
0.7644154	−13.318201	0.7675407	−13.362844	0.7689304	−13.382030

We find the ratios of the corresponding components for the 12th, 13th, 14th and 15th iterates to be

$$\begin{array}{ccc} -17.351783 & -17.378482 & -17.389552 \\ -17.393189 & -17.395694 & -17.396795 \\ -17.401310 & -17.399257 & -17.398356. \end{array}$$

The three last ratios enable us to consider λ_1 to be -17.39 or -17.40. As we have seen (§ 26), accurate to four figures, $\lambda_1 = -17.3977$.

In case the largest proper number is multiple, we have from (7), analogous to the expression we had before,

$$\frac{y_{k+1}}{y_k} = \lambda_1 + O\left(\frac{\lambda_{r+1}}{\lambda_1}\right)^k.$$

Thus in this case too, under the condition that $a_1 \neq 0$, the ratio $\dfrac{y_{k+1}}{y_k}$ gives the approximate value of the largest proper number.
The question of the multiplicity of the root cannot be solved without a more detailed investigation. We shall return to this question again in § 31.

Example 3. Let us determine the first proper number of the matrix

$$\begin{pmatrix} 1.022551 & 0.116069 & -0.287028 & -0.429969 \\ 0.228401 & 0.742521 & -0.176368 & -0.283720 \\ 0.326141 & 0.097221 & 0.197209 & -0.216487 \\ 0.433864 & 0.148965 & -0.193686 & 0.006472 \end{pmatrix}.$$

On solving the characteristic equation

$$\lambda^4 - 1.968753\lambda^3 + 1.391184\lambda^2 - 0.415291\lambda + 0.044360 = 0,$$

we obtain for the proper numbers the values

$$\lambda_1 = \lambda_2 = 0.667483, \quad \lambda_3 = 0.346148, \quad \lambda_4 = 0.287639.$$

Let us determine λ_1 by means of the iterative process, taking for the initial vector $(1, 1, 1, 1)$.

We give a table of the iterates beginning with the 9th:

$\tilde{Y}_9$	$A\tilde{Y}_9$	$\tilde{Y}_{10}$	$A\tilde{Y}_{10}$	$\tilde{Y}_{11}$	$A\tilde{Y}_{11}$
1.000000	0.666160	1.000000	0.666822	1.000000	0.667151
1.844723	1.230507	1.847165	1.232545	1.848387	1.233563
0.676506	0.449420	0.674643	0.449211	0.673660	0.449088
0.875250	0.583298	0.875613	0.584025	0.875834	0.584399
4.396479	2.929385	4.397421	2.932603	4.397881	2.934201

The ratios of the components of these iterates are computed to be

$$\begin{matrix} 0.666160 & 0.666822 & 0.667151 \\ 0.667042 & 0.667263 & 0.667373 \\ 0.664325 & 0.665850 & 0.666639 \\ 0.666466 & 0.666990 & 0.667249. \end{matrix}$$

The last four ratios give for λ_1 the value 0.667, which is correct to three places.

Let us analyse, lastly, the third case, that for which $\lambda_1 = -\lambda_2$, $|\lambda_1| > |\lambda_3|$.

From equation (6) we see that in this case the even and the odd iterates have different coefficients of corresponding powers of λ_1, since

$$y_{2k} = (c_1 + c_2)\lambda_1^{2k} + c_3\lambda_3^{2k} + \cdots + c_n\lambda_n^{2k}$$

$$y_{2k+1} = (c_1 - c_2)\lambda_1^{2k+1} + c_3\lambda_3^{2k+1} + \cdots + c_n\lambda_n^{2k+1},$$

and two neighboring iterates thus cannot be used for the determination of λ_1. We can, however, determine λ_1^2 by one of the following formulas:

$$\lambda_1^2 \simeq \frac{y_{2k+2}}{y_{2k}} \quad \text{or} \quad \lambda_1^2 \simeq \frac{y_{2k+1}}{y_{2k-1}}.$$

In all the cases we have examined, for the iterative process to be successful it is necessary that the first proper number sufficiently exceed the one next following, in point of modulus.

The process that we have described also makes possible the determination of all the components of the proper vector belonging to the largest proper number, for the ratios of the components of the vector Y_k tend to the ratios of the components of this proper vector. Indeed, for $a_1 \neq 0$:

$$Y_k = A^k Y_0 = \lambda_1^k\left[a_1 X_1 + a_2\left(\frac{\lambda_2}{\lambda_1}\right)^k X_2 + \cdots + a_n\left(\frac{\lambda_n}{\lambda_1}\right)^k X_n\right]$$

$$= a_1\lambda_1^k\left[X_1 + O\left(\frac{\lambda_2}{\lambda_1}\right)^k X_2\right].$$

If λ_1 be a proper number of multiplicity r, the indicated method makes it possible to determine one of the proper vectors belonging to λ_1. We note that in this case, by proceeding from different initial vectors, we arrive, generally speaking, at different proper vectors.

In the third case, when $\lambda_2 = -\lambda_1$, it is expedient to construct the vectors $Y_k + \lambda_1 Y_{k-1}$ and $Y_k - \lambda_1 Y_{k-1}$. The ratios of the components of these vectors will respectively tend to the ratios of the components of the vectors X_1 and X_2 belonging to the proper numbers λ_1 and λ_2.

Indeed, in view of the equation

$$Y_k = a_1\lambda_1^k X_1 + a_2(-\lambda_1)^k X_2 + a_3\lambda_3^k X_3 + \cdots$$

we have

$$Y_k + \lambda_1 Y_{k-1} = 2a_1\lambda_1^k X_1 + a_3(\lambda_3 + \lambda_1)\lambda_3^{k-1} X_3 + \cdots$$

$$= \lambda_1^k\Big(2a_1 X_1 + O\Big(\frac{\lambda_3}{\lambda_1}\Big)^k\Big),$$

$$Y_k - \lambda_1 Y_{k-1} = 2a_2(-\lambda_1)^k X_2 + a_3(\lambda_3 - \lambda_1)\lambda_3^{k-1} X_3 + \cdots$$

$$= (-\lambda_1)^k\Big(2a_2 X_2 + O\Big(\frac{\lambda_3}{\lambda_1}\Big)^k\Big).$$

As an example let us find the proper vector belonging to the first proper number of the matrix of Example 2. In § 26 we determined the components of this vector, viz.:

$$X_1 = (0.094129, -0.766896, 0.745248)$$

or, after normalizing,

$$X_1 = (1.00000, -8.14729, 7.91730).$$

From the table exhibited in Example 2, we find for the components of the vector the following values:

1.00000	1.00000	1.00000
−8.13327	−8.14133	−8.14472
7.90081	7.91026	7.91426

We see that the last result already approximates closely enough the exact value.

In concluding this section we shall find the first proper number, and the proper vector belonging to it, of the matrix

$$\begin{pmatrix} 0.22 & 0.02 & 0.12 & 0.14 \\ 0.02 & 0.14 & 0.04 & -0.06 \\ 0.12 & 0.04 & 0.28 & 0.08 \\ 0.14 & -0.06 & 0.08 & 0.26 \end{pmatrix}.$$

In Table XIII, § 17, 14 iterates were computed, starting with the vector (0.76; 0.08; 1.12; 0.68).

Computing the ratios of the components of the 14th and 13th iterates, we find for λ_1 the value 0.4800.

(We ignore the second component of the iterates because of its smallness by comparison with the other components.)

The ratios of the components of the 7th and 6th iterates give for λ_1 the values

$$0.4800 \quad 0.4792 \quad 0.4808.$$

For the components of the proper vector we find from the 14th iterate the following values:

$$(1.0000 \quad 0.0000 \quad 1.0000 \quad 1.0000).$$

It is readily verified that the exact value of $\lambda_1 = 0.48$, and that the proper vector belonging to it has the components $(1, 0, 1, 1)$.

§ 30. IMPROVING THE CONVERGENCE OF THE ITERATIVE PROCESS

The convergence of the iterative process may be very slow. The cause of this is sometimes the presence of a nonlinear elementary divisor corresponding to the first proper number. In § 34 it will be shown how to reveal this circumstance and how to proceed in such a case. But the cause of the slow convergence may also be the closeness of the second proper number to the first. In this case the convergence may be improved by means of certain devices.

1. *Raising the matrix to a power.* It is sometimes expedient to apply Lobachevsky's method to the characteristic equation of the matrix. This can be done without finding the characteristic equation in explicit form, in the following manner. Successive powers of the matrix A: $A^2, A^4, A^8, A^{16}, \ldots$ are computed. For each of these matrices the trace is computed, and then λ_1 is determined approximately by the formula

$$(1) \qquad\qquad \lambda_1 \simeq \sqrt[2^k]{\operatorname{tr} A^{2^k}}.$$

The last formula follows from the fact that

$$\operatorname{tr} A^m = \lambda_1^m + \lambda_2^m + \cdots + \lambda_n^m,$$

and consequently

$$
\begin{aligned}
\sqrt[m]{\operatorname{tr} A^m} &= \lambda_1 \sqrt[m]{1 + \left(\frac{\lambda_2}{\lambda_1}\right)^m + \cdots + \left(\frac{\lambda_n}{\lambda_1}\right)^m} \\
&= \lambda_1 + O\left(\frac{1}{m}\left(\frac{\lambda_2}{\lambda_1}\right)^m\right).
\end{aligned}
$$

(2)

Instead of extracting the root one can begin the usual iteration of an arbitrary vector, using the constructed powers, so that knowing, for instance A^8, $A^8 Y_0$ can be easily constructed, and then $A^8(A^8 Y_0)$ $=A^{16} Y_0$ and finally, $A^{17} Y_0 = A(A^{16} Y_0)$, and λ_1 then found as the ratio of the components of the vectors $A^{17} Y_0$ and $A^{16} Y_0$.

Obtaining each power of a matrix requires, however, n^3 operations, and is therefore equivalent to n iterations of the vector. If the convergence of the process of iterations is good, sufficient accuracy in the determination of λ_1 will be attained after a few steps and there will be no need to have recourse to the process described. One should turn to it, however, if the number of iterations for attaining the requisite accuracy exceeds $n \cdot \log_2 n$.

In computing the powers of matrices one should use the check described in § 25.

2. *The scalar product.* This device is particularly handy in application to a symmetric matrix; we will expound it here without that assumption, however.

Together with the sequence of iterates of the vector Y_0 by the matrix A,

(3) $Y_0, Y_1 = AY_0, Y_2 = A^2 Y_0, \ldots, Y_k = A^k Y_0, \ldots$

let there also be computed the sequence of iterates by the matrix A', the transpose of A,

(4) $Y_0, Y_1' = A' Y_0, Y_2' = A'^2 Y_0, \ldots, Y_k' = A'^k Y_0, \ldots$

Let $b_1, \ldots, b_n$ be the coordinates of the vector Y_0 with respect to the basis $X_1', \ldots, X_n'$; and $a_1, \ldots, a_n$ the coordinates of Y_0 with respect to the basis $X_1, \ldots, X_n$. We assume in addition that the bases are so selected that the system of vectors $X_1, X_2, \ldots, X_n$ and $X_1', \ldots, X_n'$ satisfies the conditions of orthogonality and normality in the sense of § 3 Paragraph 12.

Let us form the scalar product (Y_k', Y_k):

$$(Y_k', Y_k) = (A'^k Y_0, A^k Y_0) = (Y_0, A^{2k} Y_0)$$
$$= (b_1 X_1' + b_2 X_2' + \cdots + b_n X_n',$$
$$a_1 \lambda_1^{2k} X_1 + a_2 \lambda_2^{2k} X_2 + \cdots + a_n \lambda_n^{2k} X_n).$$

A further step is made possible by the properties of orthogonality and normality of the system of vectors $X_1, \ldots, X_n$ and $X_1' \ldots, X_n'$:

(5) $$(Y_k', Y_k) = a_1 b_1 \lambda_1^{2k} + a_2 b_2 \lambda_2^{2k} + \cdots + a_n b_n \lambda_n^{2k}.$$

Analogously,

(6) $$(Y_{k-1}', Y_k) = a_1 b_1 \lambda_1^{2k-1} + a_2 b_2 \lambda_2^{2k-1} + \cdots + a_n b_n \lambda_n^{2k-1}.$$

From equations (5) and (6) we obtain

(7)
$$\frac{(Y_k', Y_k)}{(Y_{k-1}', Y_k)} = \frac{a_1 b_1 \lambda_1^{2k} + a_2 b_2 \lambda_2^{2k} + \cdots + a_n b_n \lambda_n^{2k}}{a_1 b_1 \lambda_1^{2k-1} + a_2 b_2 \lambda_2^{2k-1} + \cdots + a_n b_n \lambda_n^{2k-1}}$$
$$= \lambda_1 + O\left(\frac{\lambda_2}{\lambda_1}\right)^{2k}.$$

From this estimate it is evident that the formation of the scalar product lessens the number of steps of the iteration required for the determination of λ_1 to a given accuracy by almost half. However along with this, sequence (4) must be computed in addition.

In case of a symmetric matrix, sequences (3) and (4) coincide, and thus the application of the scalar product method is particularly efficient. Beginning with a certain step of the iterations, one must compute the corresponding scalar products and determine λ_1 by their ratio, to wit:

$$\lambda_1 \simeq \frac{(A^k Y_0, A^k Y_0)}{(A^{k-1} Y_0, A^k Y_0)}.$$

Example.   Let us consider the matrix

$$\begin{pmatrix} 1.0000000 & 0 & 1.0000000 & 0 \\ 1.0000000 & 0.7777778 & 0.3333333 & 0.3333333 \\ 0 & -0.0252525 & 0.5555556 & -0.0252525 \\ 0 & -0.8888889 & -8.6444444 & 0.1111111 \end{pmatrix},$$

whose proper numbers are, 1, $\frac{2}{3}$, $\frac{4}{9}$ and $\frac{1}{3}$.

For the determination of λ_1 we form the iterates $A^k Y_0$, taking as Y_0 the vector $(1, 1, 1, 1)$.

$A^{17}Y_0$	$A^{18}Y_0$	$A^{19}Y_0$	$A^{20}Y_0$
4.6731097	4.6760089	4.6779433	4.6792336
8.3733415	8.3912886	8.4032694	8.4112637
0.0028992	0.0019344	0.0012903	0.0008605
−8.3861607	−8.3998278	−8.4089592	−8.4150555
4.6631897	4.6694041	4.6735438	4.6763023

The last row maintains a check of the computations.

The ratios of the three corresponding components of these iterates will be

1.000414   1.000276
1.001428   1.000951
1.001087   1.000725.

The last column gives $\lambda_1 \simeq 1.001$; the value found coincides with the exact one accurately to one unit in the third place.

We shall now show how this value can be refined by an application of the scalar product method.

With this in view we form iterates of the vector $(1, 1, 1, 1)$ by the transposed matrix A'.

On computing them, we obtain

$A'^{20}Y_0 = (0.7961118, -0.0002189, 3.9939022, -0.1134904)$.

Furthermore

$(A^{20}Y_0, A'^{20}Y_0) = 4.681817$ and $(A^{19}Y_0, A'^{20}Y_0) = 4.681816$.

The ratio

$$\frac{(A^{20}Y_0, A'^{20}Y_0)}{(A^{19}Y_0, A'^{20}Y_0)} = 1.000000$$

gives λ_1 the value 1.000000, correct to six decimals.

Observation.   If in finding the iterates we normalize them, then

$$\lambda_1 = \frac{(A\tilde{Y}_{k-1}, A'^{k}Y_0)}{(\tilde{Y}_{k-1}, A'^{k}Y_0)},$$

where normalization of vectors $A'^{k}Y_0$ is a matter of indifference.

3. *The δ^2 process.*[1] This device is applicable only in case $|\lambda_1| > |\lambda_2|$, and λ_1, λ_2 and λ_3 are real.

Let us assume that we have determined a number of quantities

$$(8) \qquad\qquad y_k, y_{k+1}, y_{k+2}, \ldots,$$

about which it is known that

$$(9) \qquad\qquad y_k = c_1\lambda_1^k + c_2\lambda_2^k + \cdots + c_n\lambda_n^k.$$

As y_k one can take, for instance, any component of the vector $Y_k = A^k Y_0$, the trace of the matrix A^k, the scalar product of corresponding iterates, etc.   Then, as has been shown in § 29 and § 30, the first proper number λ_1 can be approximately determined as the ratio $u_k = \dfrac{y_{k+1}}{y_k}$.

Moreover it was shown in § 29 that

$$(10) \qquad u_k = \lambda_1[1 - b_2'\alpha_2^k - b_3'\alpha_3^k + b_2 b_2'\alpha_2^{2k}] + O(\alpha_3^k + \alpha_2^{2k}),$$

where

$$b_2' = \frac{c_2}{c_1}(1-\alpha_2), \quad b_3' = \frac{c_3}{c_1}(1-\alpha_3) \quad \text{and} \quad \alpha_i = \frac{\lambda_i}{\lambda_1}.$$

If the convergence of the sequence $u_k, u_{k+1}, u_{k+2}, \ldots$ is insufficiently rapid, it can be greatly improved by the following device, which is called "the δ^2 process".

Let us form

$$(11) \qquad\qquad P(u_k) = \frac{\begin{vmatrix} u_k & u_{k+1} \\ u_{k+1} & u_{k+2} \end{vmatrix}}{u_k - 2u_{k+1} + u_{k+2}}.$$

[1] A. Aitken [1].

We shall show that

$$(12) \qquad P(u_k) = \lambda_1 + O\left(\frac{\lambda_2}{\lambda_1}\right)^{2k} + O\left(\frac{\lambda_3}{\lambda_1}\right)^k.$$

With this aim in view we put

$$u_k = \lambda_1(1 + \varepsilon_k),$$

then

$$\begin{vmatrix} u_k & u_{k+1} \\ u_{k+1} & u_{k+2} \end{vmatrix} = \lambda_1^2 \begin{vmatrix} 1 + \varepsilon_k & 1 + \varepsilon_{k+1} \\ 1 + \varepsilon_{k+1} & 1 + \varepsilon_{k+2} \end{vmatrix}.$$

Separating the last determinant into the sum of four, we obtain

$$\begin{vmatrix} u_k & u_{k+1} \\ u_{k+1} & u_{k+2} \end{vmatrix} = \lambda_1^2 \left[\varepsilon_k - 2\varepsilon_{k+1} + \varepsilon_{k+2} + \begin{vmatrix} \varepsilon_k & \varepsilon_{k+1} \\ \varepsilon_{k+1} & \varepsilon_{k+2} \end{vmatrix} \right].$$

But

$$u_k - 2u_{k+1} + u_{k+2} = \lambda_1[\varepsilon_k - 2\varepsilon_{k+1} + \varepsilon_{k+2}].$$

Thus

$$P(u_k) = \lambda_1 \left[1 + \frac{\begin{vmatrix} \varepsilon_k & \varepsilon_{k+1} \\ \varepsilon_{k+1} & \varepsilon_{k+2} \end{vmatrix}}{\varepsilon_k - 2\varepsilon_{k+1} + \varepsilon_{k+2}} \right].$$

On calculating we find that

$$\frac{\begin{vmatrix} \varepsilon_k & \varepsilon_{k+1} \\ \varepsilon_{k+1} & \varepsilon_{k+2} \end{vmatrix}}{\varepsilon_k - 2\varepsilon_{k+1} + \varepsilon_{k+2}} \simeq A\alpha_3^k + B\alpha_2^{2k},$$

where

$$A = \frac{-b_3'(\alpha_2 - \alpha_3)^2}{(1 - \alpha_2)^2}, \quad B = b_2 b_2' \alpha_2^2.$$

Thus

$$P(u_k) = \lambda_1 + O\left(\frac{\lambda_3}{\lambda_1}\right)^k + O\left(\frac{\lambda_2}{\lambda_1}\right)^{2k}.$$

Hence it follows that the error in the determination of λ_1 with the aid of the δ^2 process may be less by far than in a direct determination of it from the sequence $u_k, u_{k+1}, \ldots$.

Observation. In finding the first proper number of a symmetric matrix, the δ^2 process must be applied not to the components of the iterates, but to corresponding scalar products.

Lastly we note a property of the operation $P(u_k)$ that is convenient for practical computations, viz.:

$$(13) \qquad\qquad P(u_k+c) = c+P(u_k).$$

Hence it follows that before employing the δ^2 process, one can deduct a convenient constant from the numbers u_k, u_{k+1}, u_{k+2}. (Such a constant could be, for example, one consisting of the number of decimals that have already been established.)

We shall show the application of the δ^2 process to the examples of § 29.

In Example 2, § 29, for the ratios displayed the application of the δ^2 process gives for λ_1 the following values:

$$-17.3974$$
$$-17.3977$$
$$-17.3977.$$

Accurate to four decimals, $\lambda_1 = -17.3977$.

Analogously, in Example 3, § 29, the application of the δ^2 process to the displayed ratios gives for λ_1 the following values:

$$0.66748$$
$$0.66748$$
$$0.66748$$
$$0.66748.$$

Thus λ_1 is already determined with an accuracy of five decimals (the exact value of $\lambda_1 = 0.667483$).

The δ^2 process may also be applied to the determination of the components of the first proper vector. In this connection we shall show two distinct variants of this process, the choice depending on whether λ_1 can be considered as known to a sufficient degree of accuracy or not.

1) Let there be known to us only the sequence of iterates $A^k Y_0$, where for computational convenience let each iterate be normalized by division by a fixed component z_k.

We shall denote any other component of the vector Y_k by y_k.

If

$$y_k = c_1\lambda_1^k + c_2\lambda_2^k + \cdots + c_n\lambda_n^k$$

$$z_k = b_1\lambda_1^k + b_2\lambda_2^k + \cdots + b_n\lambda_n^k,$$

then the indicated division reduces z_k to unity and y_k to v_k, where

$$v_k = \frac{y_k}{z_k} = \frac{c_1\lambda_1^k + c_2\lambda_2^k + \cdots + c_n\lambda_n^k}{b_1\lambda_1^k + b_2\lambda_2^k + \cdots + b_n\lambda_n^k}$$

$$= \frac{c_1}{b_1} + \frac{c_2 b_1 - b_2 c_1}{b_1^2}\left(\frac{\lambda_2}{\lambda_1}\right)^k + O\left(\frac{\lambda_3}{\lambda_1}\right)^k.$$

A simple computation shows that

$$P(v_k) = \frac{c_1}{b_1} + O\left(\frac{\lambda_3}{\lambda_1}\right)^k + O\left(\frac{\lambda_2}{\lambda_1}\right)^{2k}.$$

Thus if the δ^2 process is applied to all components of the normalized vector $A^k Y_0$, we shall find the ratios of the coefficients standing with the powers of λ_1 in the expressions for y_k. These coefficients, moreover, are proportional to the components of the proper vector.

Thus for the proper vector of Example 2 of the preceding section, application of the δ^2 process gives for the components of the vector the values

$$\begin{pmatrix} 1.00000 \\ -8.14718 \\ 7.91721 \end{pmatrix},$$

which are considerably closer to the exact ones than the values computed directly from the same iterates.

2) If λ_1 is known with sufficient exactitude, an improving process can be constructed in the following manner. We multiply all the components of the vectors $A^{k-1}Y_0$, $A^k Y_0$, $A^{k+1}Y_0$, by λ_1, 1, λ_1^{-1} respectively, and then apply the δ^2 process to them. Since

$$y_{k-1}\lambda_1 = \lambda_1^k c_1 + \lambda_2^{k-1}\lambda_1 c_2 + \cdots + \lambda_n^{k-1}\lambda_1 c_n$$

$$y_k = \lambda_1^k c_1 + \lambda_2^k c_2 + \cdots + \lambda_n^k c_n$$

$$y_{k+1}\lambda_1^{-1} = \lambda_1^k c_1 + \frac{\lambda_2^{k+1}}{\lambda_1} c_2 + \cdots + \frac{\lambda_n^{k+1}}{\lambda_1} c_n,$$

we have

$$\begin{vmatrix} \lambda_1 y_{k-1} & y_k \\ y_k & \lambda_1^{-1} y_{k+1} \end{vmatrix} = [c_1 c_2 \lambda_1^{k-1} \lambda_2^{k-1} (\lambda_1 - \lambda_2)^2 + \cdots$$

$$+ c_1 c_n \lambda_1^{k-1} \lambda_n^{k-1} (\lambda_1 - \lambda_n)^2] \left[1 + O\left(\frac{\lambda_3}{\lambda_1}\right)^k \right].$$

Moreover

$$\lambda_1 y_{k-1} - 2 y_k + \lambda^{-1} y_{k+1} = c_2 \frac{\lambda_2^{k-1}}{\lambda_1} (\lambda_1 - \lambda_2)^2 + \cdots + c_n \frac{\lambda_n^{k-1}}{\lambda_1} (\lambda_1 - \lambda_n)^2.$$

Hence

$$\frac{\begin{vmatrix} \lambda_1 y_{k-1} & y_k \\ y_k & \lambda_1^{-1} y_{k+1} \end{vmatrix}}{y_{k-1} - 2 y_k + y_{k+1}} = c_1 \lambda_1^k \left[1 + O\left(\frac{\lambda_3}{\lambda_1}\right)^k \right].$$

The ratios of the numbers obtained, which latter are computed for the different components, give the ratios of the components of the proper vector.

§ 31. FINDING THE PROPER NUMBERS NEXT IN LINE

In this section we shall show that by modifying the process of iteration in a certain manner, one can determine the product of several proper numbers. This opens up for us the possibility of determining one after another the proper numbers next in order after the first. It should be remarked, however, that the determination of the product of even two of the proper numbers encounters great difficulties, chief of which is the disappearance of significant figures. As a rule, therefore, even the second proper number determined by means of the iterative process is of a much lesser degree of accuracy than the first.

Let

(1)
$$y_k = c_1 \lambda_1^k + c_2 \lambda_2^k + \cdots + c_n \lambda_n^k$$
$$z_k = d_1 \lambda_1^k + d_2 \lambda_2^k + \cdots + d_n \lambda_n^k,$$

be any components of the vectors $A^k Y_0$ and $A^k Z_0$. (If $Z_0 = Y_0$,

y_k and z_k will be any two components whatever of the vector $A^k Y_0$.)
Let us form the expression

$$(2) \qquad [y_k, z_k] = \begin{vmatrix} y_k & y_{k+1} \\ z_k & z_{k+1} \end{vmatrix}.$$

On working this out, we find that

$$(3) \qquad [y_k, z_k] = (c_2 d_1 - c_1 d_2)(\lambda_1 - \lambda_2)\lambda_1^k \lambda_2^k$$
$$+ (c_3 d_1 - c_1 d_3)(\lambda_1 - \lambda_3)\lambda_1^k \lambda_3^k + O(\lambda_2^k \lambda_3^k).$$

Hence it follows that

$$(4) \qquad \frac{[y_{k+1}, z_{k+1}]}{[y_k, z_k]} = \lambda_1 \lambda_2 + O\left(\frac{\lambda_3}{\lambda_1}\right)^k.$$

In particular, one may take $Z_0 = A Y_0$, whereupon

$$(5) \qquad [y_k, z_k] = [y_k, y_{k+1}] = \begin{vmatrix} y_k & y_{k+1} \\ y_{k+1} & y_{k+2} \end{vmatrix}.$$

It can be verified that

$$(6) \qquad [y_k, y_{k+1}] = c_1 c_2 (\lambda_1 - \lambda_2)^2 \lambda_1^k \lambda_2^k + c_1 c_3 (\lambda_1 - \lambda_3)^2 \lambda_1^k \lambda_3^k + O(\lambda_2^k \lambda_3^k).$$

Therefore

$$(7) \qquad \frac{[y_{k+1}, y_{k+2}]}{[y_k, y_{k+1}]} = \lambda_1 \lambda_2 + O\left(\frac{\lambda_3}{\lambda_1}\right)^k.$$

The disappearance of significant figures which we alluded to is occasioned by the fact that the elements of determinant (2) and of determinant (5) as well become almost proportional to one another to the degree that the step of the iteration is increased.

Observation. In case the first proper number is multiple, $\lambda_1 = \lambda_2 = \cdots = \lambda_r$, but the elementary divisors corresponding to it are linear, the process indicated gives the product $\lambda_1 \lambda_{r+1}$. This process will generally give the product of the two "eldest" roots of the minimum, and not of the characteristic, polynomial. Therefore in utilizing the iterative process, one can judge of the multiplicity of the dominant proper number only by means of indirect considerations. An indication of the order of the multiplicity can be given, for example, by a comparison of the trace of the matrix with the computed magnitudes of the first and second roots of the minimum polynomial.

It is obvious that by means of three components of the vector $A^k Y_0$ or by means of corresponding components of the vectors $A^k Y_0$, $A^{k+1} Y_0$, $A^{k+2} Y_0$, we can analogously construct determinants $[y_k, z_k, t_k]$ or $[y_k, y_{k+1}, y_{k+2}]$. The ratios of corresponding determinants will give us an approximate value for the product $\lambda_1 \lambda_2 \lambda_3$. The process can, theoretically, be continued to the determination of the product of the k roots. Practically, the process yields only very rough values for the product of two or three roots.

As an example we shall determine the second proper number for the matrix we examined in Paragraph 2, § 30.

Adopting for y_k the second component of the vector $A^k X_0$, and for z_k the fourth component, we obtain

$$[y_{18}, z_{18}] = \begin{vmatrix} 8.3912886 & 8.4032694 \\ -8.3998278 & -8.4089592 \end{vmatrix} = 0.0240124.$$

Analogously

$$[y_{19}, z_{19}] = 0.0159949.$$

Thus $\lambda_1 \lambda_2 \simeq 0.666110$.

In § 30 we saw that $\lambda_1 \simeq 1.001$. This gives the value $\lambda_2 = 0.6654$. If we use the value refined by the scalar product method, $\lambda_1 = 1.000000$, we will obtain as λ_2 the value 0.666, which is correct to an accuracy of one unit in the third decimal:

$$(\lambda_2 = \tfrac{2}{3} = 0.66666 \ldots).$$

If as y_k the first component of the vector $A^k X_0$ be adopted, and the second component as z_k, we will obtain

$$[y_{18}, z_{18}] = 0.0397902$$
$$[y_{19}, z_{19}] = 0.0265541.$$

This gives

$$\lambda_1 \lambda_2 \simeq 0.667353.$$

Utilizing the refined value of λ_1, we obtain $\lambda_2 \simeq 0.667$. Finally we shall determine λ_2 by the ratio of determinants of the form

$$\begin{vmatrix} y_k & y_{k+1} \\ y_{k+1} & y_{k+2} \end{vmatrix},$$

where y_k is any component of the vector $A^k X_0$. Taking as y_k the first component, we obtain

$$[y_{18}, y_{19}] = \begin{vmatrix} 4.6760089 & 4.6779433 \\ 4.6779433 & 4.6792336 \end{vmatrix} = -0.0030156.$$

Analogously,

$$[y_{17}, y_{18}] = -0.0045170.$$

Thus $\lambda_1 \lambda_2 \simeq 0.66761$ and $\lambda_2 \simeq 0.668$.

§ 32. DETERMINATION OF THE PROPER NUMBERS NEXT IN LINE AND THEIR PROPER VECTORS AS WELL

The process of determining the proper numbers following after λ_1 described by us in the preceding section, does not make possible the determination of the proper vectors belonging to these numbers. In this section we shall give devices that permit the determination not only of the proper numbers that follow λ_1, but also of the proper vectors belonging to them.

1. *The λ-difference.* Let there be computed the sequence

$$(1) \qquad y_1, y_2, \ldots, y_m, \ldots, y_k, \ldots$$

and from it let $\lambda_1 = \dfrac{y_{k+1}}{y_k}$ be determined. Here y_k is any component of the iterate $Y_k = A^k Y_0$.

Let us form the difference

$$(2) \qquad y_{k+1} - \lambda_1 y_k = c_2(\lambda_2 - \lambda_1)\lambda_2^k + \ldots + c_n(\lambda_n - \lambda_1)\lambda_n^k.$$

If λ_2 is greater in modulus than all the rest of the proper numbers and $c_2 \neq 0$, the first term of this difference will predominate and we shall be able to determine λ_2 in a way analogous to that by which we determined λ_1, viz.:

$$(3) \qquad \lambda_2 \simeq \frac{y_{k+1} - \lambda_1 y_k}{y_k - \lambda_1 y_{k-1}}.$$

However in such a determination of λ_2 we encounter a disappearance of significant figures, since in the numerator and denominator of ratio (3) we have to subtract quantities close to one another. It is expedient in practice, after finding λ_1 from the ratio of y_{k+1} and y_k, to turn back and determine λ_2 from the ratio

$$(4) \qquad \lambda_2 \simeq \frac{y_{m+1} - \lambda_1 y_m}{y_m - \lambda_1 y_{m-1}}, \quad m < k,$$

taking as m the least number for which the predominance of λ_2 over the succeeding proper numbers has already begun to make itself felt. The suggested device gives for λ_2 values that are rough, to be sure; however they are frequently adequate for practical purposes. Theoretically it is possible by means of an analogous process to determine the succeeding proper numbers too.

It is obvious that to determine the second proper vector the process of forming the λ-difference must be carried out on the sequence $AY_0, A^2Y_0, \ldots, A^kY_0, \ldots$; indeed, the difference

$$A^{k+1}Y_0 - \lambda_1 A^k Y_0 = a_2(\lambda_2 - \lambda_1)\lambda_2^k X_2 + \cdots + a_n(\lambda_n - \lambda_1)\lambda_n^k X_n$$

shows that the components of the vector X_2 may be found in a manner analogous to that with which we determined the components of the vector X_1 in § 29.

For an example, we determine the second proper number of the matrix discussed in § 30, Paragraph 2.

As λ_1 we shall take not only the value obtained directly from the ratios of the components of the 20th and 19th iterates ($\lambda_1 \simeq 1.001$), but also the value refined by means of the scalar product ($\lambda_1 \simeq 1.000000$).

Adopting as y_k the first component of the vector $A^k Y_0$, we obtain (for $\lambda_1 \simeq 1.000000$), taking into account the 17th, 18th and 19th iterates ($m = 18$):

$$\lambda_2 \simeq \frac{y_{m+1} - \lambda_1 y_m}{y_m - \lambda_1 y_{m-1}} \simeq \frac{4.677943 - 4.676009}{4.676009 - 4.673110} = \frac{0.001934}{0.002899} = 0.6671.$$

Analogously, adopting as y_k the fourth component of the vector $A^k Y_0$, we obtain

$$\lambda_2 \simeq \frac{-0.009131}{-0.013667} \simeq 0.6681.$$

Thus knowledge of a sufficiently exact value of λ_1 has made it possible to determine λ_2 also with sufficient accuracy (to three places after the decimal).   (Exactly, $\lambda_2 = 0.666 \ldots$.)

If as λ_1 we take the rougher value $\lambda_1 \simeq 1.001$, on computing the previous ratio we run up against the phenomenon described—the disappearance of significant figures.   In this case as m a number considerably less than 20 must be taken.

Thus, considering the 9th, 10th and 11th iterates of the vector $A^k Y_0$,

$A^9 Y_0$	$A^{10} Y_0$	$A^{11} Y_0$
4.4665336	4.5365193	4.5841480
7.1243407	7.5407651	7.8281626
0.0699857	0.0476287	0.0321941
-7.4707539	-7.7678185	-7.9777169
4.1901061	4.3570946	4.4667878,

we obtain, on computing the values $y_{m+1} - \lambda_1 y_m$:

$m = 9$	$m = 10$
0.06552	0.04309
0.40930	0.27986
-0.02242	-0.01548
-0.28959	$-0.20213.$

The ratios of these quantitites give for λ_2 the values

$$0.658$$
$$0.684$$
$$0.690$$
$$0.698.$$

Thus knowledge of a value for λ_1 that is very rough allows us nonetheless to obtain for λ_2 a value accurate to three units of the second place, by using the early iterates.

The λ-difference method permits the determination of the components of the second proper vector too.

For the matrix under consideration, the second proper vector has the components

$$(1, \tfrac{31}{5} = 6.2, -\tfrac{1}{3} = -0.333\ldots, -\tfrac{71}{15} = -4.733\ldots).$$

Approximate values for the components of the second proper vector can be obtained as the respective ratios of the components of the vector $A^{m+1}Y_0 - \lambda_1 A^m Y_0$. Taking $m = 9$, we obtain, by utilizing the components of the vector $A^{10}Y_0 - \lambda_1 A^9 Y_0$, computed earlier, the following values for the components of the proper vector (after normalization):

$$(1.00; \ 6.49; \ -0.36; \ -4.69),$$

which is in sufficiently close agreement with the exact value.

2. *The method of exhaustion.* The method of exhaustion in the general case requires not only knowledge of the first proper number, but also of the proper vectors that correspond to it for the matrix A and its transpose as well.

Let us say that we have determined with sufficient accuracy λ_1, X_1 and X_1', where the vectors X_1 and X_1' have been so normalized that their scalar product is equal to unity.

Regarding the components of the vector X_1 as a column, and the components of the vector X_1' as a row, we form the matrix product $X_1 X_1'$. This will be the square matrix

$$\begin{pmatrix} x_{11}x_{11}' & x_{11}x_{21}' & \cdots & x_{11}x_{n1}' \\ x_{21}x_{11}' & x_{21}x_{21}' & \cdots & x_{21}x_{n1}' \\ \cdot & \cdot & \cdots & \cdot \\ x_{n1}x_{11}' & x_{n1}x_{21}' & \cdots & x_{n1}x_{n1}' \end{pmatrix}.$$

We remark that the matrix product $X_1' X_1$ is a number, to wit the scalar product of the vectors X_1' and X_1.

Next form the matrix

(5) $$A_1 = A - \lambda_1 X_1 X_1'.$$

We shall show that the matrix A_1 has the same proper numbers and vectors as the matrix A, with the exception of the first proper number, in place of which there will be a proper number equal to zero. Indeed,

(6)
$$A_1 X_1 = A X_1 - \lambda_1 (X_1 X_1') X_1$$
$$= A X_1 - \lambda_1 X_1 (X_1' X_1) = A X_1 - \lambda_1 X_1 = 0$$
$$A_1 X_i = A X_i - \lambda_1 (X_1 X_1') X_i$$
$$= A X_i - \lambda_1 X_1 (X_1' X_i) = \lambda_i X_i, \quad (i \neq 1)$$

since $X_1' X_1 = 1$, and $X_1' X_i = 0$, on the strength of the orthogonal properties of the vectors $X_1, \ldots, X_n$ and $X_1', \ldots, X_n'$.

The property of the matrix A_1 which we have indicated makes it possible to start from the vector sequence $A_1 Y_0, \ldots, A_1{}^m Y_0, \ldots$ and determine λ_2 and X_2 in a manner analogous to that by which we determined λ_1 and X_1 from the sequence $A Y_0, \ldots, A^k Y_0, \ldots$, since the proper number λ_2 will be the first proper number of the matrix A_1. We shall call this process the "process of exhaustion".

We shall show that

(7)
$$A_1^m Y_0 = A^m Y_0 - \lambda_1^m X_1 X_1' Y_0,$$

i.e., that in the practical application of the indicated process there is no need of computing the matrix A_1 in fact, or of forming the series of vectors $A_1 Y_0, \ldots, A_1{}^m Y_0, \ldots$, but that it is sufficient to compute just two adjacent vectors $A_1{}^{m+1} Y_0$ and $A_1{}^m Y_0$ by formula (7).

In order to establish equation (7) we shall introduce the so-called *bilinear resolution of the matrix A*.

On the basis of the orthonormal properties of the system of proper vectors of the matrix A and its transpose, the following matrix equation is valid:

$$I = X_1 X_1' + X_2 X_2' + \cdots + X_n X_n'.$$

Multiplying this equation on the left by A and replacing $A X_i$ by $\lambda_i X_i$, $i = 1, \ldots, n$, we obtain

(8)
$$A = \lambda_1 X_1 X_1' + \lambda_2 X_2 X_2' + \cdots + \lambda_n X_n X_n'.$$

The process of exhaustion annihilates the first summand in this resolution, whence it follows that

$$A_1 = \lambda_2 X_2 X'_2 + \cdots + \lambda_n X_n X'_n.$$

Moreover

$$A^m = \lambda_1^m X_1 X'_1 + \lambda_2^m X_2 X'_2 + \cdots + \lambda_n^m X_n X'_n.$$

Analogously

$$A_1^m = \lambda_2^m X_2 X'_2 + \cdots + \lambda_n^m X_n X'_n.$$

Therefore

$$A_1^m = A^m - \lambda_1^m X_1 X'_1,$$

whence issues equation (7).

Thus for the practical application of the process of exhaustion, one has to compute the vector $X_1 X'_1 Y_0$, form the vectors $A_1^{m+1} Y_0$ and $A_1^m Y_0$ by formula (7), and then find λ_2 and X_2 in a manner analogous to that by which we determined λ_1 and X_1 in § 29. It is obvious that in course of this all the devices for improving the convergence of the iterations which we have described in § 30 may be employed, in particular the δ^2 process.

We note that one must take as m a number considerably less than the number of the iterate from which the components of the first proper vector are determined. We note too that in case one finds it necessary to determine with great accuracy the second proper number and the vector belonging to it, one will find it in fact expedient to form the matrix A_1 and to compute its iterates.

Example. Let us again consider the matrix of § 30, Paragraph 2.

The method of exhaustion requires, for the determination of the second proper number, knowledge of the first proper number, as also of the proper vectors of the matrix A and of matrix A' as well that belong to it. It is therefore necessary in utilizing this method to compute, along with the sequence of iterates $A^k Y_0$, that of $A'^k Y_0$ also. Thus in determining λ_1 we must always refine the value obtained, by means of the scalar product method.

In our example, using twelve iterations of the vector $Y_0 = (1, 1, 1, 1)$ by the matrix A and by A', we have obtained for λ_1 the value $\lambda_1 = 1.000000$ (see § 30, Paragraph 2).

For the components of the proper vectors of the matrices A and A'

we obtain, after normalizing the components of the vectors $A^{20}Y_0$ and $A'^{20}Y_0$, the values

$$
\begin{array}{rr}
1.00000 & 1.00000 \\
1.79757 & -0.00027 \\
0.00018 & 5.01676 \\
-1.79838 & -0.14256.
\end{array}
$$

(The exact values of the components of the first proper vector of the matrix A are $(1, 1.8, 0, -1.8)$.) Following the theory, it is first necessary to normalize the vectors X_1 and X'_1 so that $(X_1, X'_1) = 1$. The normalizing factor computes out to be $c = 0.795678$. Thus for the components of the first proper vectors of the matrices A and A' we obtain the values

$$
\begin{array}{rr}
1.00000 & 0.79568 \\
1.79757 & -0.00021 \\
0.00018 & 3.99173 \\
-1.79838 & -0.11343.
\end{array}
$$

Now we can form the matrix product $X_1 X'_1$; it is:

$$
X_1 X'_1 = \begin{pmatrix}
0.79568 & -0.00021 & 3.99173 & -0.11343 \\
1.43029 & -0.00038 & 7.17541 & -0.20390 \\
0.00014 & 0 & 0.00072 & -0.00002 \\
-1.43093 & 0.00038 & -7.17865 & 0.20399
\end{pmatrix}.
$$

Next we form the matrix A_1:

$$
A_1 = A - \lambda X_1 X'_1
$$

$$
= \begin{pmatrix}
0.20432 & 0.00021 & -2.99173 & 0.11343 \\
-0.43029 & 0.77816 & -6.84208 & 0.53723 \\
-0.00014 & -0.02525 & 0.55484 & -0.02523 \\
1.43093 & -0.88927 & -1.46579 & -0.09288
\end{pmatrix},
$$

and form the iterates of the vector $Y_0 = (1, 1, 1, 1)$ by this matrix.

We adduce the 17th and 18th iterates of the matrix A_1:

$A_1^{17} Y_0$	$A_1^{18} Y_0$
-0.00869	-0.00580
-0.05388	-0.03597
0.00290	0.00193
0.04107	0.02741
-0.01861	-0.01242 .

The ratios of the components of the 17th and 18th iterates give for λ_2 the values

$$0.667$$
$$0.668$$
$$0.666$$
$$0.667 \, .$$

The ratios of the corresponding components of the 18th iterate give for the components of the second proper vector the values

$$1.00$$
$$6.20$$
$$-0.333$$
$$-4.73 \quad .$$

We see that not only the second proper number λ_2 but also the components of the second proper vector are determined more exactly by the method of exhaustion than by the λ-difference method. However this method requires, for the case of a nonsymmetric matrix, much additional work. In the case of a symmetric matrix the method of exhaustion can be recommended.

In computing the second proper number and the components of the second proper vector, one can utilize the modification of the process that was described, whereby the vector $A_1^k Y_0$ is computed, but the iterations of the matrix A_1 are avoided, by the formula

$$A_1^k Y_0 = A^k Y_0 - \lambda_1^k X_1 X_1' Y_0.$$

Making the computation, we obtain

$$X_1 X_1' Y_0 = \begin{pmatrix} 4.67377 \\ 8.40142 \\ 0.00084 \\ -8.40521 \end{pmatrix}.$$

Now we compute $A_1^k Y_0$ for $k=9$ and 10. This gives

$k=9$	$k=10$
-0.20724	-0.13725
-1.27708	-0.86065
0.06915	0.04679
0.93446	0.63739 .

Hence we find λ_2:

$$0.662$$
$$0.674$$
$$0.677$$
$$0.682$$

and from $A_1^{10} Y_0$, the components of the proper vector: $(1.00, 6.27, -0.34, -4.64)$.

3. *The reduction method.* Given the matrix A, let the first proper number λ_1 and the proper vector $X_1 = (x_1, \ldots, x_n)$ belonging to it have been computed. Let us consider the matrix

$$P = \begin{pmatrix} x_1 & 0 & \ldots & 0 \\ x_2 & 1 & \ldots & 0 \\ \cdot & \cdot & \cdot & \cdot \cdot \cdot \\ x_n & 0 & \ldots & 1 \end{pmatrix}.$$

We confirm without difficulty that

$$
P^{-1} = \begin{pmatrix}
\dfrac{1}{x_1} & 0 & \dots & 0 \\[2ex]
-\dfrac{x_2}{x_1} & 1 & \dots & 0 \\[1ex]
\cdot & \cdot & \cdot & \cdot \\[1ex]
-\dfrac{x_n}{x_1} & 0 & \dots & 1
\end{pmatrix}.
$$

The matrix $P^{-1}AP$ is similar to the matrix A, and the proper numbers of both matrices are thus identical.

But

$$
P^{-1}AP = \begin{pmatrix}
\lambda_1 & \dfrac{a_{12}}{x_1} & \dots & \dfrac{a_{1n}}{x_1} \\[2ex]
0 & a_{22} - \dfrac{x_2}{x_1} a_{12} & \dots & a_{2n} - \dfrac{x_2}{x_1} a_{1n} \\[1ex]
\cdot & \cdot \cdot \cdot \cdot \cdot & & \cdot \\[1ex]
0 & a_{n2} - \dfrac{x_n}{x_1} a_{12} & \dots & a_{nn} - \dfrac{x_n}{x_1} a_{1n}
\end{pmatrix}
$$

$$
= \begin{pmatrix}
\lambda_1 & b_{12} & \dots & b_{1n} \\
0 & & & \\
\vdots & & B & \\
0 & & &
\end{pmatrix}.
$$

Thus

$$
|P^{-1}AP - \lambda I| = (\lambda_1 - \lambda)|B - \lambda I|.
$$

Consequently the proper numbers of the matrix A will be λ_1 and the proper numbers of B, a matrix of the $(n-1)$th order. If X_1 be so normalized that $x_1 = 1$, we will have

$$
B = \begin{pmatrix}
a_{22} - x_2 a_{12} & \dots & a_{2n} - x_2 a_{1n} \\
\cdot \cdot \cdot & \cdot \cdot \cdot \cdot & \cdot \cdot \cdot \\
a_{n2} - x_n a_{12} & \dots & a_{nn} - x_n a_{1n}
\end{pmatrix}.
$$

For the determination of λ_2 we must obviously construct a sequence $BY_0, \ldots, B^kY_0, \ldots$ and find λ_2 as the ratio of any components of the constructed vectors.

Furthermore let Y be a proper vector of the matrix B. The matrix $P^{-1}AP$ will then have a proper vector $\begin{pmatrix} y_1 \\ Y \end{pmatrix}$. Let us determine y_1.

We have

$$
\begin{pmatrix} \lambda_1 & a_{12} & \cdots & a_{1n} \\ 0 & & & \\ & & B & \\ 0 & & & \end{pmatrix} \begin{pmatrix} y_1 \\ Y \end{pmatrix}
$$

$$
= \begin{pmatrix} \lambda_1 y_1 + a_{12}y_2 + \cdots + a_{1n}y_n \\ BY \end{pmatrix} = \lambda \begin{pmatrix} y_1 \\ Y \end{pmatrix}.
$$

Equating the first components of this vector equation, we obtain

$$
\lambda_1 y_1 + a_{12}y_2 + \cdots + a_{1n}y_n = \lambda y_1,
$$

whence

$$
y_1 = \frac{a_{12}y_2 + \cdots + a_{1n}y_n}{\lambda - \lambda_1}.
$$

Finally the proper vector X_i of the matrix A is determined by the formula

$$
X_i = P\begin{pmatrix} y_1^{(i)} \\ Y_i \end{pmatrix} = \begin{pmatrix} y_1^{(i)} \\ x_2 y_1^{(i)} + y_2^{(i)} \\ \cdots \cdots \cdots \\ x_n y_1^{(i)} + y_n^{(i)} \end{pmatrix}.
$$

Example. Let us determine the second proper number and the components of the proper vector belonging to it, for the matrix of § 30, Paragraph 2.

As we have seen, the first proper number for this matrix has been determined to be $\lambda_1 \simeq 1.001$ (§ 30, Paragraph 2), and the components of the proper vector belonging to it to be $(1, 1.79757, 0.00018, -1.79838)$ (§ 33).

For determining λ_2 we compute the matrix B:

$$B = \begin{pmatrix} 0.77778 & -1.46424 & 0.33333 \\ -0.02525 & 0.55538 & -0.02525 \\ -0.88889 & -6.84606 & 0.11111 \end{pmatrix}.$$

Next we form iterates of the vector $Y_0 = (1, 1, 1)$ by the matrix B. We give the result of the 15th and 16th iterations:

$B^{15}X_0$	$B^{16}X_0$
−0.09565	−0.06388
0.00725	0.00484
0.06339	0.04243
−0.02502	−0.01661 .

We determine λ_2 from the ratios of the components of the 16th and 15th iterates. This gives

$$0.668$$
$$0.668$$
$$0.669 .$$

We now determine the components of the second proper vector. With this in view, we determine, to begin with,

$$y_1 = \frac{a_{12}y_2 + a_{13}y_3 + a_{14}y_4}{\lambda_2 - \lambda_1} = \frac{0.00484}{-0.333} \simeq -0.0145.$$

Next we compute the components x_2, x_3, x_4 of vector X_2:

$$x_2 = -0.0900$$
$$x_3 = 0.00484$$
$$x_4 = 0.0685.$$

Thus the components of the vector X_2 are $(-0.0145, -0.0900, 0.00484, 0.0685)$, or, after normalization, $(1, 6.21, -0.333, -4.72)$.

As we see, the reduction method makes possible the exact determination not only of the second proper number itself but also of the proper vector belonging to it.

§ 33. DETERMINATION OF THE FIRST PROPER NUMBER. SECOND CASE

Let us now consider the case where the elementary divisors of the matrix are linear, but the first proper number λ_1 is complex. In this case $\bar{\lambda}_1$ will also be a proper number, so that of the proper numbers that are of greatest modulus there will be not less than two. Let us assume that $|\lambda_3| < |\lambda_1|$. Let X_1 and X_2 be the proper vectors belonging to the proper numbers λ_1 and λ_2. On the strength of § 3, Paragraph 6, one may consider their components to be complex conjugates. Furthermore, if

$$Y_0 = a_1 X_1 + a_2 X_2 + \cdots + a_n X_n$$

is a real initial vector, a_1 and a_2 will be complex conjugate numbers.

Reasoning just as in § 29, we form the sequence of vectors

(1) $$AY_0, \ldots, A^k Y_0, \ldots \quad .$$

Any component of the vector $Y_k = A^k Y_0$ will obviously have its previous form,

(2) $$y_k = c_1 \lambda_1^k + c_2 \lambda_2^k + \cdots + c_n \lambda_n^k,$$

where c_1 and c_2 are complex conjugates. Let

(3) $$c_1 = Re^{i\alpha}, \quad c_2 = Re^{-i\alpha},$$
$$\lambda_1 = re^{i\theta}, \quad \lambda_2 = re^{-i\theta} ;$$

then

(4) $$y_k = 2Rr^k \cos (k\theta + \alpha) + c_3 \lambda_3^k + \cdots \quad .$$

The presence of the factor $\cos(k\theta + \alpha)$ will be the cause of a marked oscillation of the values of y_k, in magnitude and in sign as well. Thus the presence of complex roots that are the largest in modulus will at once be revealed in forming the iterates. Since $r^2 = \lambda_1 \lambda_2$, we can determine the quantity r^2 by means of the determinants introduced in § 31. Here, as distinct from the case of real roots, we will not encounter the disappearance of significant figures, so that r^2 is determined with sufficient accuracy from the ratio of these determinants.

We shall show that having computed r, one can determine $\cos \theta$, utilizing the sequence $y_1, y_2, \ldots, y_k, \ldots$ for this purpose.

We introduce the expression[1]

$$(5) \qquad \mu_k = \tfrac{1}{2}[ry_{k-1} + r^{-1}y_{k+1}].$$

We have from (4), accurately to terms of the order of $\left(\dfrac{\lambda_3}{\lambda_1}\right)^k$, that

$$y_k \simeq 2Rr^k \cos(k\theta + \alpha).$$

Furthermore, since

$$\cos[(k+1)\theta + \alpha] + \cos[(k-1)\theta + \alpha] = 2\cos(k\theta + \alpha)\cos\theta,$$

we obviously have

$$(6) \qquad \cos\theta \simeq \frac{\mu_k}{y_k},$$

with an accuracy to quantities of the order of $\left(\dfrac{\lambda_3}{\lambda_1}\right)^k$.

Since the case of complex roots is of little practical interest, we shall not dwell on the determination of the proper vector. For the same reason an illustrative example is not given. We have touched upon this case as upon the cases considered in the preceding section, only in order to show the possible peculiarities of the course of the iterative process.

§ 34. THE CASE OF A MATRIX WITH NON-LINEAR ELEMENTARY DIVISORS

In all the preceding sections devoted to iterative methods, it has been assumed that the elementary divisors of the matrix are linear, i.e., that the matrix A is reducible to diagonal form. In this section we abandon that requirement. As has already been noted in the introduction, the iterative process depends essentially on the structure of the Jordan canonical form that is connected with the given matrix. We shall not consider the general case, but shall show in a most simple example only the character of the changes that occur if the matrix has nonlinear elementary divisors.

---

[1] A. Aitken [1].

236 *Proper Numbers and Proper Vectors of a Matrix*

Let us concern ourselves with the determination of the first proper number λ_1 of the matrix A. We shall discuss the case in which λ_1 is real and belongs to the box $\begin{pmatrix} \lambda_1 & 0 \\ 1 & \lambda_1 \end{pmatrix}$ in the Jordan canonical form, and the next proper number λ_2 is less than λ_1 in modulus. To simplify the calculation we shall consider the elementary divisors that correspond to the rest of the proper numbers to be linear.

As previously, we adopt an arbitrary vector Y_0 and form the sequence $AY_0, A^2Y_0, \ldots, A^kY_0, \ldots$. We shall elucidate how any component y_k of the vector $Y_k = A^kY_0$ depends on the corresponding powers of λ_i in such a case. Since the proper vectors of the matrix do not in this case form a basis, let us take as $X_1, \ldots, X_n$ the vectors of that basis relative to which the linear transformation with the matrix A is brought into canonical form. Then

$$AX_1 = \lambda_1 X_1 + X_2$$
$$AX_2 = \lambda_1 X_2$$
(1) $$AX_3 = \lambda_3 X_3$$
$$\cdots \cdots \cdots \cdots$$
$$AX_n = \lambda_n X_n.$$

Let

(2) $$X_j = (x_{1j}, x_{2j}, \ldots, x_{nj}), \quad Y_k = (y_{1k}, \ldots, y_{nk}),$$

and

$$Y_0 = a_1 X_1 + a_2 X_2 + \cdots + a_n X_n.$$

We have, on the strength of (1):

(4) $$AY_0 = a_1\lambda_1 X_1 + a_1 X_2 + a_2\lambda_1 X_2 + a_3\lambda_3 X_3 + \cdots + a_n\lambda_n X_n$$
$$\cdots \cdots \cdots \cdots \cdots \cdots$$

(5) $$A^kY_0 = a_1\lambda_1^k X_1 + a_1 k\lambda_1^{k-1} X_2 + a_2\lambda_1^k X_2 + a_3\lambda_3^k X_3 + \cdots + a_n\lambda_n^k X_n.$$

Furthermore, in view of (2), we have

$$y_{ik} = a_1\lambda_1^k x_{i1} + (a_1\lambda_1^{k-1}k + a_2\lambda_1^k)x_{i2} + a_3\lambda_3^k x_{i3} + \cdots + a_n\lambda_n^k x_{in}$$
$$= (a_1 x_{i1} + a_2 x_{i2})\lambda_1^k + a_1 x_{i2} k\lambda_1^{k-1} + a_3 x_{i3}\lambda_3^k + \cdots + a_n x_{in}\lambda_n^k.$$

Thus any component of the vector Y_k will have the form (we omit the first index, as before):

$$(6) \qquad y_k = c_1\lambda_1^k + c_2 k\lambda_1^{k-1} + c_3\lambda_3^k + \cdots + c_n\lambda_n^k.$$

The ratio $\dfrac{y_{k+1}}{y_k}$ tends to λ_1 as before, but more slowly than any geometrical progression, owing to the presence of the factor k in the second term, to wit:

$$\frac{y_{k+1}}{y_k} = \lambda_1\left[1 + O\left(\frac{1}{k}\right)\right].$$

To determine λ_1 from the ratio $\dfrac{y_{k+1}}{y_k}$ becomes practically impossible.

We note that if the box to which λ_1 belongs in the Jordan canonical form has a more complex structure, other powers of λ_1 also appear in expression (6); they are multiplied by the corresponding binomial coefficients:

$$y_k = c_1\lambda_1^k + c_2 k\lambda_1^{k-1} + c_3\frac{k(k-1)}{2}\lambda_1^{k-2} + \cdots + c_n\lambda_n^k.$$

The ratio $\dfrac{y_{k+1}}{y_k}$ approaches λ_1 still more slowly.

In general a slow convergence of the process provides grounds for surmising that a nonlinear elementary divisor, connected with the largest proper number, is present.

However the process by means of which the product of two proper numbers is determined will converge rapidly enough in the case in hand. It can be confirmed without difficulty that

$$(7) \qquad \frac{[y_{k+1},\, y_{k+2}]}{[y_k,\, y_{k+1}]} = \lambda_1^2 + O\left(\frac{\lambda_3}{\lambda_1}\right)^k,$$

for indeed on computing $[y_k,\, y_{k+1}] = \begin{vmatrix} y_k & y_{k+1} \\ y_{k+1} & y_{k+2} \end{vmatrix}$ we find that $[y_k,\, y_{k+1}] = -c_2^2\lambda_1^{2k} + O\left(\frac{\lambda_3}{\lambda_1}\right)^k$, whence (7) follows.

Ratio (7) makes possible the determination of λ_1.

We shall now show how, in case λ_1 belongs to the box $\begin{pmatrix} \lambda_1 & 0 \\ 1 & \lambda_1 \end{pmatrix}$ and this circumstance is suspected from the slow convergence of the sequence y_k, one can satisfy oneself that this is indeed the fact.

Let us have found λ_1 from the ratio (7).   Let us construct

$$
\begin{aligned}
\Delta y_k = y_{k+1} - \lambda_1 y_k &= c_1 \lambda_1^{k+1} + c_2 (k+1) \lambda_1^k + c_3 \lambda_3^{k+1} + \cdots \\
&\quad + c_n \lambda_n^{k+1} - c_1 \lambda_1^{k+1} - c_2 k \lambda_1^k - c_3 \lambda_3^k \lambda_1 - \cdots - c_n \lambda_n^k \lambda_1 \\
&= c_2 \lambda_1^k + O(\lambda_3^k).
\end{aligned}
$$

(8)

Then

$$
\frac{\Delta y_{k+1}}{\Delta y_k} = \lambda_1 + O\left(\frac{\lambda_3}{\lambda_1}\right)^k,
$$

i.e., $\dfrac{\Delta y_{k+1}}{\Delta y_k}$ tends to λ_1 fast enough.   The coincidence of the limit of $\dfrac{\Delta y_{k+1}}{\Delta y_k}$ with the value for λ_1 computed earlier, and the fact of the rapid convergence of $\dfrac{\Delta y_{k+1}}{\Delta y_k}$ to λ_1 serve as confirmation of the surmise that λ_1 figures in a box of the second order.

Instead of the computation described, one can have recourse to computation of the second λ-difference.

If λ_2 belongs to a box of the second order, the second λ-difference

$$
\Delta^2 y_k = \Delta y_{k+1} - \lambda_1 \Delta y_k = O(\lambda_3^k)
$$

will be small in comparison with the component y_k itself.

The case of nonlinear elementary divisors of the matrix is very rarely met with in practice.   We shall therefore not dwell either on the determination of the proper vector that belongs to the first proper number, or on the determination of the succeeding proper numbers and the vectors belonging to them.

We remark only that all the devices we have discussed for the case of linear elementary divisors may be employed in the general case too, after suitable modification.[1]

[1] A. Aitken [1]; K. A. Semendiaev [1].

§ 35. IMPROVING THE CONVERGENCE OF THE ITERATIVE PROCESS FOR SOLVING SYSTEMS OF LINEAR EQUATIONS [1]

As we saw in § 17, in solving a system of linear equations given in the form $X = AX + F$ by the method of iteration, we consider that $X \simeq X^{(k)} = AX^{(k-1)} + F$. Here

$$(1) \qquad X - X^{(k)} = \left(X^{(k+1)} - X^{(k)} \right) + \left(X^{(k+2)} - X^{(k+1)} \right) + \cdots$$

Let us calculate approximately the sum of this series.
We have

$$(2) \qquad X^{(k+1)} - X^{(k)} = A\left(X^{(k)} - X^{(k-1)} \right) = \cdots = A^k\left(X^{(1)} - X^{(0)} \right).$$

Let $\lambda_1, \ldots, \lambda_n$ be the proper numbers of the matrix A; we shall consider them to be distinct, for simplicity, and to be arranged in order of diminishing moduli, and shall consider $U_1, U_2, \ldots, U_n$ to be the proper vectors that correspond to them. Let us resolve the vector $X^{(1)} - X^{(0)}$ in terms of the proper vectors of the matrix A:

$$X^{(1)} - X^{(0)} = \alpha_1 U_1 + \cdots + \alpha_n U_n.$$

Then, on the strength of (2):

$$X^{(k+1)} - X^{(k)} = \alpha_1 \lambda_1^k U_1 + \cdots + \alpha_n \lambda_n^k U_n = \alpha_1 \lambda_1^k U_1 + O(\lambda_2^k).$$

As was shown in § 29, we can determine the first proper number λ_1 from the ratio of the components of the vectors $(X^{(k+1)} - X^{(k)})$ and $(X^{(k)} - X^{(k-1)})$, with sufficiently large k. Moreover

$$\left(X^{(k+1)} - X^{(k)} \right) + \left(X^{(k+2)} - X^{(k+1)} \right) + \cdots$$

$$(3) \quad = \alpha_1 \lambda_1^k (1 + \lambda_1 + \lambda_1^2 + \cdots) U_1 + \cdots + \alpha_n \lambda_n^k (1 + \lambda_n + \lambda_n^2 + \cdots) U_n$$

$$= \frac{\alpha_1 \lambda_1^k}{1 - \lambda_1} U_1 + \cdots + \frac{\alpha_n \lambda_n^k}{1 - \lambda_n} U_n \simeq \frac{X^{(k+1)} - X^{(k)}}{1 - \lambda_1} + O(\lambda_2^k).$$

We have the right to sum the geometrical progressions, for all $|\lambda_i| < 1$, since we consider the iterative process to be convergent.

[1] L. A. Liusternik [1].

Thus, accurate to quantities of the order of λ_2^k, we have, as the solution of the linear system, the expression

(4) $$X \simeq X^{(k)} + \frac{X^{(k+1)} - X^{(k)}}{1 - \lambda_1}.$$

The extra term in equation (4) is effortlessly computed, and greatly improves the convergence of the iterative process.

If the solution of the linear system is approximately determined by means of the expression $X \simeq \sum\limits_{l=1}^{k} A^l F$, then, reasoning analogously, we find that

$$X \simeq \sum_{l=1}^{k} A^l F + \frac{A^{k+1} F}{1 - \lambda_1}$$

will give a better approximation to the sought solution. Here the quantity λ_1 is found approximately from the ratio of the components of the vectors $A^{k+1} F$ and $A^k F$.

Thus for the example analysed by us in § 17, we obtained (see § 29), on proceeding from the 14th approximation, $\lambda_1 = 0.4800$.

Furthermore, from Table XIII, § 17, we have

$$X^{(13)} = (1.53490847,\ 0.12200958,\ 1.97509985,\ 1.41289889).$$

On computing $\dfrac{A^{k+1} F}{1 - \lambda_1}$, we obtain

$$\frac{A^{k-1} F}{1 - \lambda_1} = (0.00005656,\ 0.00005656,\ 0.00005656).$$

Thus

$$X \simeq (1.534965,\ 0.122010,\ 1.975156,\ 1.412955).$$

We see that, with accuracy to the sixth figure, the improved solution coincides with that found by Gauss's method.

Moreover,

$$\sum_{l=0}^{6} A^l F = (1.52533,\ 0.12201,\ 1.96551,\ 1.40333).$$

Having taken $\lambda_1 = 0.4792$ (see § 29) and used the device for improvement we find that

$$\sum_{l=0}^{6} A^l F + \frac{A^7 F}{0.5208} = (1.53495, 0.12203, 1.97514, 1.41293).$$

As we see from Table XIII of § 17, the approximation obtained coincides better with the exact solution than does the 14th approximation.

The method for improving the convergence here suggested may also be applied to the Seidel method, for that method, applied to the system $X = AX + F$, is equivalent to application of the usual iterative method to the system

$$X = A_1 X + F, \text{ where } A_1 = (I - B)^{-1} C.$$

Here B and C are the triangular constituent matrices of the matrix A (see § 19). The largest proper number μ_1 of the matrix A_1 is determined from the ratio of the components of the vectors $X^{(k+1)} - X^{(k)}$ and $X^{(k)} - X^{(k-1)}$, and for improving the convergence one can as before utilize the formula

$$X \approx X^{(k)} + \frac{X^{(k+1)} - X^{(k)}}{1 - \mu_1}.$$

In conclusion we make mention of methods that have quite recently appeared in the literature, which are based on the ideas of functional analysis. These methods, developed for the solution of more general problems of operational calculus, give, in particular, a solution of the basic problems of linear algebra (not only the solution of a linear algebraic system, but also the finding of the proper numbers of a matrix), if for the operator under study one adopts the linear transformation connected with the given matrix.

The reader who is interested in these methods and who is familiar with the basic ideas of functional analysis is referred to the extensive article of L. V. Kantorovich [1], which contains a large bibliography, and also to another work of the same author, [2], as well as to those of I. P. Natanson [1], M. Sh. Birman [1], and M. K. Gavurin [1].

BIBLIOGRAPHY

AITKEN, A. C.   [1] *Studies in practical mathematics.*   II: *The evaluation of the latent roots and latent vectors of a matrix.*   Proc. Roy. Soc. Edinburgh, **57** (1937); 269–304.

—— [2] *Determinants and matrices.*   New York, 1948 (Interscience Pub.)

BANACHIEWICZ, TADEUSZ.   [1] *Principes d'une nouvelle technique de la méthode des moindres carrés.*   Comptes-rendus mensuels des sciences mathématiques et naturelles, Akademija umiejętności, Krakow, Jan., 1938.

BIRMAN, M. Sh.   [1] *Some estimates for the method of steepest descent.*   (Russian) Uspekhi matem. nauk, **5** (1950), No. 3; 152–155.

DANILEVSKIĬ, A. M.   [1] *The numerical solution of the secular equation.*   (Russian) Matem. sbornik, **44** (1937), No. 2; 169–171.

DWYER, PAUL S.   [1] *The solution of simultaneous equations.*   Psychometrika, **6** (1941); 101–129.

FADDEEV, D. K., AND SOMINSKIĬ, I. S.   [1] *Sbornik zadach po vysshei algebry (Collection of problems on higher algebra).*   2nd ed., Moscow, 1949.   (Gostekhizdat.)

FADDEEV, D. K.   [2] *On the transformation of the secular equation of a matrix.*   (Russian) Leningrad: Trudy inst. inzh. prom. stroit., **4** (1937); 78–86.

FRAZER, R. A., DUNCAN, W. J., AND COLLAR, A. R.   [1] *Elementary matrices and some applications to dynamic and differential equations.*   Cambridge, 1938.   (Cambridge University Press.)

GANTMAKHER, F. R.   [1] *Trudy 2-go Vsesoiuznogo Matem. S"ezda (Works of the Second All-Union Mathematical Congress),* **2** (1936), 45–48.

GAVURIN, M. K.   [1] *Application of the polynomials of best approximation to the improve ment of the convergence of iterative processes.*   (Russian) Uspekhi matem. nauk, **5** (1950), No. 3; 156–160.

GEL'FAND, I. M.   [1] *Lektsii po lineinoi algebry (Lectures on linear algebra).*   2nd ed., Moscow, 1951.   (Gostekhizdat.)

HOTELLING, H.   [1] *Some new methods in matrix calculation.*   Ann. Math. Stat., **4** (1943); 1–33.

KANTOROVICH, L. V.   [1] *Functional analysis and applied mathematics.*   (Russian) Uspekhi matem. nauk, **3** (1948), No. 6; 89–185.

KANTOROVICH, L. V. [2] *On Newton's method.* (Russian) Trudy matem. inst. im. V. A. Steklova, **28** (1949); 104–144.

KHLODOVSKIĬ, I. N. [1] *On the theory of the general case of transforming the secular equation by Krylov's method.* Izvestiîa akad. nauk SSSR, Otdel matem. i estest. nauk, Ser. 7 (1933); 1077–1102.

KRYLOV, A. N. [1] *Über die numerische Auflösung einer Gleichung, durch die in technischen Fragen die Frequenz kleiner Schwingungen bestimmt ist.* Izvestiîa akad. nauk SSSR, Otdel matem. i estest. nauk, Ser. 7 (1931), No. 4; 491–539. (Russian.)

KUROSH, A. G. [1] *Kurs vysshei algebry (Course of higher algebra).* 2nd ed., Moscow, 1950. (Gostekhizdat.)

LEVERRIER, U. J. J. [1] Ann. l'observ. Paris. T. 11, 128 ff.

—— [2] *Sur les variations séculaire des éléments des orbites pour les sept planètes principales.* J. de Math., s. 1 (1840), **5**, 230 ff.

LIUSTERNIK, L. A. [1] *Remarks on the numerical solution of boundary-value problems for Laplace's equation and the calculation of characteristic values by the method of nets.* (Russian) Trudy matem. inst. im. V. A. Steklova, **20** (1947); 49–64.

LUZIN, N. N. [1] *Sur la méthode de Mr. A. Krylov de composition de l'équation séculaire.* (Russian) Izvestiîa akad. nauk SSSR, Otdel matem. i estest. nauk, Ser. 7 (1931); 903–958.

—— [2] *Sur certaines propriétés du multiplicateur inversant dans le procédé de Mr. A. Krylov.* (Russian) Izvestiîa akad. nauk SSSR, Otdel matem. i estest. nauk, Ser. 7 (1932); I: 595–638; II: 735–762; III: 1065–1102.

MAL'TSEV, A. N. [1] *Osnovy lineĭnoĭ algebry (Foundations of linear algebra).* Moscow, 1948. (Gostekhizdat.)

MEHMKE, R., AND NEKRASOV, P. A. [1] *Auflösung eines lineares systems von Gleichungen durch successive Annäherung.* (German and Russian) Matem. sbornik, **16** (1892); 437–459.

MIKELADZE, SH. E. [1] *On the evaluation of determinants whose elements are polynomials.* (Russian) Prikladnaîa matem. i mekh., **12** (1948), No. 2; 219–222.

MORRIS, JOSEPH. [1] *An escalator process for the solution of linear simultaneous equations.* Philos. Mag. **37** (1946), No. 7; 106–110.

—— [2] *The escalator process for the solution of damped Lagrangian frequency equations.* Philos. Mag., **38** (1947), No. 7; 275–287.

MORRIS, JOSEPH, and HEAD, J. W. [1] *The "escalator" process for the solution of Lagrangian frequency equations.* Philos. Mag. **35** (1944), No. 7; 735–759.

NATANSON, I. P. [1] Uchënye zapiski Len. Ped. inst. im. Gertsena, **64** (1948). (Russian.)

NEKRASOV, P. A. [1] *Die Bestimmung der Unbekannten nach der Methode der kleinsten Quadrate bei einer sehr grossen Anzahl der Unbekannten.* (Russian) Matem. sbornik, **12** (1885); 189–204.

NIKOLAEVA, M. V. [1] *On Southwell's relaxation method.* (Russian) Trudy matem. inst. im. V. A. Steklova, **28** (1949); 160–182.

REICH, EDGAR. [1] *On the convergence of the classical iterative method of solving linear simultaneous equations.* Ann. Math. Stat., **20** (1949), No. 3; 448–451.

Bibliography

SAMUELSON, P. [1] *A method of determining explicitly the coefficients of the characteristic equation.* Ann. Math. Stat., **13** (1942); 424–429.

SEMENDIAEV, K. A. [1] *The determination of the latent roots and the invariant manifolds of matrices by means of iterations.* (Russian) Priklad. matem. i mekh. **7** (1943); 193–222.

SMIRNOV, V. I. [1] *Kurs vyssheĭ matematiki (Course of higher mathematics).* Vol. 3, Part 1, Moscow, 1949. (Gostekhizdat.)

UMANSKIĬ, A. A. [1] *Kurs stroitel'noĭ mekhaniki (Course of structural mechanics).* 1935.

TURNBULL, H. W., AND AITKEN, A. C. [1] *An introduction to the theory of canonical matrices.* London, 1948. (Blackie and Son.)

WAUGH, F. V., AND DWYER, P. S. [1] *Compact computation of the inverse of a matrix.* Ann. Math. Stat., **16** (1945); 259–271.

WAYLAND, HAROLD. [1] *Expansion of the determinantal equation into polynomial form.* Q. Appl. Math., **2** (1945); 277–306.

INDEX

(Numbers refer to pages; those in **bold-face type** identify computational tables.)

SOME DOVER SCIENCE BOOKS

SOME DOVER SCIENCE BOOKS

WHAT IS SCIENCE?,
Norman Campbell

This excellent introduction explains scientific method, role of mathematics, types of scientific laws. Contents: 2 aspects of science, science & nature, laws of science, discovery of laws, explanation of laws, measurement & numerical laws, applications of science. 192pp. 5⅜ x 8.   60043-2 Paperbound $1.25

FADS AND FALLACIES IN THE NAME OF SCIENCE,
Martin Gardner

Examines various cults, quack systems, frauds, delusions which at various times have masqueraded as science. Accounts of hollow-earth fanatics like Symmes; Velikovsky and. wandering planets; Hoerbiger; Bellamy and the theory of multiple moons; Charles Fort; dowsing, pseudoscientific methods for finding water, ores, oil. Sections on naturopathy, iridiagnosis, zone therapy, food fads, etc. Analytical accounts of Wilhelm Reich and orgone sex energy; L. Ron Hubbard and Dianetics; A. Korzybski and General Semantics; many others. Brought up to date to include Bridey Murphy, others. Not just a collection of anecdotes, but a fair, reasoned appraisal of eccentric theory. Formerly titled *In the Name of Science.* Preface. Index. x + 384pp. 5⅜ x 8.
20394-8 Paperbound $2.00

PHYSICS, THE PIONEER SCIENCE,
L. W. Taylor

First thorough text to place all important physical phenomena in cultural-historical framework; remains best work of its kind. Exposition of physical laws, theories developed chronologically, with great historical, illustrative experiments diagrammed, described, worked out mathematically. Excellent physics text for self-study as well as class work. Vol. 1: Heat, Sound: motion, acceleration, gravitation, conservation of energy, heat engines, rotation, heat, mechanical energy, etc. 211 illus. 407pp. 5⅜ x 8. Vol. 2: Light, Electricity: images, lenses, prisms, magnetism, Ohm's law, dynamos, telegraph, quantum theory, decline of mechanical view of nature, etc. Bibliography. 13 table appendix. Index. 551 illus. 2 color plates. 508pp. 5⅜ x 8.
60565-5, 60566-3 Two volume set, paperbound $5.50

THE EVOLUTION OF SCIENTIFIC THOUGHT FROM NEWTON TO EINSTEIN,
A. d'Abro

Einstein's special and general theories of relativity, with their historical implications, are analyzed in non-technical terms. Excellent accounts of the contributions of Newton, Riemann, Weyl, Planck, Eddington, Maxwell, Lorentz and others are treated in terms of space and time, equations of electromagnetics, finiteness of the universe, methodology of science. 21 diagrams. 482pp. 5⅜ x 8.
20002-7 Paperbound $2.50

A SOURCE BOOK IN MATHEMATICS,
D. E. Smith
Great discoveries in math, from Renaissance to end of 19th century, in English translation. Read announcements by Dedekind, Gauss, Delamain, Pascal, Fermat, Newton, Abel, Lobachevsky, Bolyai, Riemann, De Moivre, Legendre, Laplace, others of discoveries about imaginary numbers, number congruence, slide rule, equations, symbolism, cubic algebraic equations, non-Euclidean forms of geometry, calculus, function theory, quaternions, etc. Succinct selections from 125 different treatises, articles, most unavailable elsewhere in English. Each article preceded by biographical introduction. Vol. I: Fields of Number, Algebra. Index. 32 illus. 338pp. 5⅜ x 8. Vol. II: Fields of Geometry, Probability, Calculus, Functions, Quaternions. 83 illus. 432pp. 5⅜ x 8.
60552-3, 60553-1 Two volume set, paperbound $5.00

FOUNDATIONS OF PHYSICS,
R. B. Lindsay & H. Margenau
Excellent bridge between semi-popular works & technical treatises. A discussion of methods of physical description, construction of theory; valuable for physicist with elementary calculus who is interested in ideas that give meaning to data, tools of modern physics. Contents include symbolism; mathematical equations; space & time foundations of mechanics; probability; physics & continua; electron theory; special & general relativity; quantum mechanics; causality. "Thorough and yet not overdetailed. Unreservedly recommended," *Nature* (London). Unabridged, corrected edition. List of recommended readings. 35 illustrations. xi + 537pp. 5⅜ x 8.
60377-6 Paperbound $3.50

FUNDAMENTAL FORMULAS OF PHYSICS,
ed. by D. H. Menzel
High useful, full, inexpensive reference and study text, ranging from simple to highly sophisticated operations. Mathematics integrated into text—each chapter stands as short textbook of field represented. Vol. 1: Statistics, Physical Constants, Special Theory of Relativity, Hydrodynamics, Aerodynamics, Boundary Value Problems in Math, Physics, Viscosity, Electromagnetic Theory, etc. Vol. 2: Sound, Acoustics, Geometrical Optics, Electron Optics, High-Energy Phenomena, Magnetism, Biophysics, much more. Index. Total of 800pp. 5⅜ x 8.
60595-7, 60596-5 Two volume set, paperbound $4.75

THEORETICAL PHYSICS,
A. S. Kompaneyets
One of the very few thorough studies of the subject in this price range. Provides advanced students with a comprehensive theoretical background. Especially strong on recent experimentation and developments in quantum theory. Contents: Mechanics (Generalized Coordinates, Lagrange's Equation, Collision of Particles, etc.), Electrodynamics (Vector Analysis, Maxwell's equations, Transmission of Signals, Theory of Relativity, etc.), Quantum Mechanics (the Inadequacy of Classical Mechanics, the Wave Equation, Motion in a Central Field, Quantum Theory of Radiation, Quantum Theories of Dispersion and Scattering, etc.), and Statistical Physics (Equilibrium Distribution of Molecules in an Ideal Gas, Boltzmann Statistics, Bose and Fermi Distribution, Thermodynamic Quantities, etc.). Revised to 1961. Translated by George Yankovsky, authorized by Kompaneyets. 137 exercises. 56 figures. 529pp. 5⅜ x 8½.
60972-3 Paperbound $3.50

AN INTRODUCTION TO THE GEOMETRY OF N DIMENSIONS,
D. H. Y. Sommerville
An introduction presupposing no prior knowledge of the field, the only book
in English devoted exclusively to higher dimensional geometry. Discusses
fundamental ideas of incidence, parallelism, perpendicularity, angles between
linear space; enumerative geometry; analytical geometry from projective and
metric points of view; polytopes; elementary ideas in analysis situs; content of
hyper-spacial figures. Bibliography. Index. 60 diagrams. 1966pp. 5⅜ x 8.
60494-2 Paperbound $1.50

ELEMENTARY CONCEPTS OF TOPOLOGY, *P. Alexandroff*
First English translation of the famous brief introduction to topology for the
beginner or for the mathematician not undertaking extensive study. This unusually useful intuitive approach deals primarily with the concepts of complex,
cycle, and homology, and is wholly consistent with current investigations.
Ranges from basic concepts of set-theoretic topology to the concept of Betti
groups. "Glowing example of harmony between intuition and thought," David
Hilbert. Translated by A. E. Farley. Introduction by D. Hilbert. Index. 25
figures. 73pp. 5⅜ x 8. 60747-X Paperbound $1.25

ELEMENTS OF NON-EUCLIDEAN GEOMETRY,
D. M. Y. Sommerville
Unique in proceeding step-by-step, in the manner of traditional geometry.
Enables the student with only a good knowledge of high school algebra and
geometry to grasp elementary hyperbolic, elliptic, analytic non-Euclidean geom-
etries; space curvature and its philosophical implications; theory of radical
axes; homothetic centres and systems of circles; parataxy and parallelism;
absolute measure; Gauss' proof of the defect area theorem; geodesic representation; much more, all with exceptional clarity. 126 problems at chapter endings
provide progressive practice and familiarity. 133 figures. Index. xvi + 274pp.
5⅜ x 8. 60460-8 Paperbound $2.00

INTRODUCTION TO THE THEORY OF NUMBERS, *L. E. Dickson*
Thorough, comprehensive approach with adequate coverage of classical literature, an introductory volume beginners can follow. Chapters on divisibility,
congruences, quadratic residues & reciprocity. Diophantine equations, etc. Full
treatment of binary quadratic forms without usual restriction to integral coefficients. Covers infinitude of primes, least residues. Fermat's theorem. Euler's
phi function, Legendre's symbol, Gauss's lemma, automorphs, reduced forms,
recent theorems of Thue & Siegel, many more. Much material not readily
available elsewhere. 239 problems. Index. I figure. viii + 183pp. 5⅜ x 8.
60342-3 Paperbound $1.75

MATHEMATICAL TABLES AND FORMULAS,
compiled by Robert D. Carmichael and Edwin R. Smith
Valuable collection for students, etc. Contains all tables necessary in college
algebra and trigonometry, such as five-place common logarithms, logarithmic
sines and tangents of small angles, logarithmic trigonometric functions, natural
trigonometric functions, four-place antilogarithms, tables for changing from
sexagesimal to circular and from circular to sexagesimal measure of angles, etc.
Also many tables and formulas not ordinarily accessible, including powers,
roots, and reciprocals, exponential and hyperbolic functions, ten-place logarithms of prime numbers, and formulas and theorems from analytical and
elementary geometry and from calculus. Explanatory introduction. viii +
269pp. 5⅜ x 8½. 60111-0 Paperbound $1.50

CELESTIAL OBJECTS FOR COMMON TELESCOPES,
Rev. T. W. Webb
Classic handbook for the use and pleasure of the amateur astronomer. Of
inestimable aid in locating and identifying thousands of celestial objects. Vol I,
The Solar System: discussions of the principle and operation of the telescope,
procedures of observations and telescope-photography, spectroscopy, etc., precise
location information of sun, moon, planets, meteors. Vol. II, The Stars:
alphabetical listing of constellations, information on double stars, clusters, stars
with unusual spectra, variables, and nebulae, etc. Nearly 4,000 objects noted.
Edited and extensively revised by Margaret W. Mayall, director of the American
Assn. of Variable Star Observers. New Index by Mrs. Mayall giving the location
of all objects mentioned in the text for Epoch 2000. New Precession Table
added. New appendices on the planetary satellites, constellation names and
abbreviations, and solar system data. Total of 46 illustrations. Total of xxxix
+ 606pp. 5⅜ x 8.   20917-2, 20918-0 Two volume set, paperbound $5.00

PLANETARY THEORY,
E. W. Brown and C. A. Shook
Provides a clear presentation of basic methods for calculating planetary orbits
for today's astronomer. Begins with a careful exposition of specialized mathematical topics essential for handling perturbation theory and then goes on to
indicate how most of the previous methods reduce ultimately to two general
calculation methods: obtaining expressions either for the coordinates of planetary positions or for the elements which determine the perturbed paths. An
example of each is given and worked in detail. Corrected edition. Preface.
Appendix. Index. xii + 302pp. 5⅜ x 8½.   61133-7 Paperbound $2.25

STAR NAMES AND THEIR MEANINGS,
Richard Hinckley Allen
An unusual book documenting the various attributions of names to the
individual stars over the centuries. Here is a treasure-house of information on
a topic not normally delved into even by professional astronomers; provides a
fascinating background to the stars in folk-lore, literary references, ancient
writings, star catalogs and maps over the centuries. Constellation-by-constella-
tion analysis covers hundreds of stars and other asterisms, including the
Pleiades, Hyades, Andromedan Nebula, etc. Introduction. Indices. List of
authors and authorities. xx + 563pp. 5⅜ x 8½.   21079-0 Paperbound $3.00

A SHORT HISTORY OF ASTRONOMY, *A. Berry*
Popular standard work for over 50 years, this thorough and accurate volume
covers the science from primitive times to the end of the 19th century. After
the Greeks and the Middle Ages, individual chapters analyze Copernicus, Brahe,
Galileo, Kepler, and Newton, and the mixed reception of their discoveries.
Post-Newtonian achievements are then discussed in unusual detail: Halley,
Bradley, Lagrange, Laplace, Herschel, Bessel, etc. 2 Indexes. 104 illustrations,
9 portraits. xxxi + 440pp. 5⅜ x 8.   20210-0 Paperbound $2.75

SOME THEORY OF SAMPLING, *W. E. Deming*
The purpose of this book is to make sampling techniques understandable to
and useable by social scientists, industrial managers, and natural scientists
who are finding statistics increasingly part of their work. Over 200 exercises,
plus dozens of actual applications. 61 tables. 90 figs. xix + 602pp. 5⅜ x 8½.
        61755-6 Paperbound $3.50

CHANCE, LUCK AND STATISTICS: THE SCIENCE OF CHANCE,
Horace C. Levinson
Theory of probability and science of statistics in simple, non-technical language. Part I deals with theory of probability, covering odd superstitions in regard to "luck," the meaning of betting odds, the law of mathematical expectation, gambling, and applications in poker, roulette, lotteries, dice, bridge, and other games of chance. Part II discusses the misuse of statistics, the concept of statistical probabilities, normal and skew frequency distributions, and statistics applied to various fields—birth rates, stock speculation, insurance rates, advertising, etc. "Presented in an easy humorous style which I consider the best kind of expository writing," Prof. A. C. Cohen, Industry Quality Control. Enlarged revised edition. Formerly titled *The Science of Chance*. Preface and two new appendices by the author. xiv + 365pp. 5⅜ x 8. 21007-3 Paperbound $2.00

BASIC ELECTRONICS,
prepared by the U.S. Navy Training Publications Center
A thorough and comprehensive manual on the fundamentals of electronics. Written clearly, it is equally useful for self-study or course work for those with a knowledge of the principles of basic electricity. Partial contents: Operating Principles of the Electron Tube; Introduction to Transistors; Power Supplies for Electronic Equipment; Tuned Circuits; Electron-Tube Amplifiers; Audio Power Amplifiers; Oscillators; Transmitters; Transmission Lines; Antennas and Propagation; Introduction to Computers; and related topics. Appendix. Index. Hundreds of illustrations and diagrams. vi + 471pp. 6½ x 9¼.
61076-4 Paperbound $2.95

BASIC THEORY AND APPLICATION OF TRANSISTORS,
prepared by the U.S. Department of the Army
An introductory manual prepared for an army training program. One of the finest available surveys of theory and application of transistor design and operation. Minimal knowledge of physics and theory of electron tubes required. Suitable for textbook use, course supplement, or home study. Chapters: Introduction; fundamental theory of transistors; transistor amplifier fundamentals; parameters, equivalent circuits, and characteristic curves; bias stabilization; transistor analysis and comparison using characteristic curves and charts; audio amplifiers; tuned amplifiers; wide-band amplifiers; oscillators; pulse and switching circuits; modulation, mixing, and demodulation; and additional semiconductor devices. Unabridged, corrected edition. 240 schematic drawings, photographs, wiring diagrams, etc. 2 Appendices. Glossary. Index. 263pp. 6½ x 9¼. 60380-6 Paperbound $1.75

GUIDE TO THE LITERATURE OF MATHEMATICS AND PHYSICS,
N. G. Parke III
Over 5000 entries included under approximately 120 major subject headings of selected most important books, monographs, periodicals, articles in English, plus important works in German, French, Italian, Spanish, Russian (many recently available works). Covers every branch of physics, math, related engineering. Includes author, title, edition, publisher, place, date, number of volumes, number of pages. A 40-page introduction on the basic problems of research and study provides useful information on the organization and use of libraries, the psychology of learning, etc. This reference work will save you hours of time. 2nd revised edition. Indices of authors, subjects, 464pp. 5⅜ x 8.
60447-0 Paperbound $2.75

COLLEGE ALGEBRA, H. B. Fine
Standard college text that gives a systematic and deductive structure to algebra; comprehensive, connected, with emphasis on theory. Discusses the commutative, associative, and distributive laws of number in unusual detail, and goes on with undetermined coefficients, quadratic equations, progressions, logarithms, permutations, probability, power series, and much more. Still most valuable elementary-intermediate text on the science and structure of algebra. Index. 1560 problems, all with answers. x + 631pp. 5⅜ x 8. 60211-7 Paperbound $2.75

HIGHER MATHEMATICS FOR STUDENTS OF CHEMISTRY AND PHYSICS, J. W. Mellor
Not abstract, but practical, building its problems out of familiar laboratory material, this covers differential calculus, coordinate, analytical geometry, functions, integral calculus, infinite series, numerical equations, differential equations, Fourier's theorem, probability, theory of errors, calculus of variations, determinants. "If the reader is not familiar with this book, it will repay him to examine it," *Chem. & Engineering News.* 800 problems. 189 figures. Bibliography. xxi + 641pp. 5⅜ x 8. 60193-5 Paperbound $3.50

TRIGONOMETRY REFRESHER FOR TECHNICAL MEN, A. A. Klaf
A modern question and answer text on plane and spherical trigonometry. Part I covers plane trigonometry: angles, quadrants, trigonometrical functions, graphical representation, interpolation, equations, logarithms, solution of triangles, slide rules, etc. Part II discusses applications to navigation, surveying, elasticity, architecture, and engineering. Small angles, periodic functions, vectors, polar coordinates, De Moivre's theorem, fully covered. Part III is devoted to spherical trigonometry and the solution of spherical triangles, with applications to terrestrial and astronomical problems. Special time-savers for numerical calculation. 913 questions answered for you! 1738 problems; answers to odd numbers. 494 figures. 14 pages of functions, formulae. Index. x + 629pp. 5⅜ x 8. 20371-9 Paperbound $3.00

CALCULUS REFRESHER FOR TECHNICAL MEN, A. A. Klaf
Not an ordinary textbook but a unique refresher for engineers, technicians, and students. An examination of the most important aspects of differential and integral calculus by means of 756 key questions. Part I covers simple differential calculus: constants, variables, functions, increments, derivatives, logarithms, curvature, etc. Part II treats fundamental concepts of integration: inspection, substitution, transformation, reduction, areas and volumes, mean value, successive and partial integration, double and triple integration. Stresses practical aspects! A 50 page section gives applications to civil and nautical engineering, electricity, stress and strain, elasticity, industrial engineering, and similar fields. 756 questions answered. 556 problems; solutions to odd numbers. 36 pages of constants, formulae. Index. v + 431pp. 5⅜ x 8. 20370-0 Paperbound $2.25

INTRODUCTION TO THE THEORY OF GROUPS OF FINITE ORDER, R. Carmichael
Examines fundamental theorems and their application. Beginning with sets, systems, permutations, etc., it progresses in easy stages through important types of groups: Abelian, prime power, permutation, etc. Except 1 chapter where matrices are desirable, no higher math needed. 783 exercises, problems. Index. xvi + 447pp. 5⅜ x 8. 60300-8 Paperbound $3.00

FIVE VOLUME "THEORY OF FUNCTIONS" SET BY KONRAD KNOPP

This five-volume set, prepared by Konrad Knopp, provides a complete and readily followed account of theory of functions. Proofs are given concisely, yet without sacrifice of completeness or rigor. These volumes are used as texts by such universities as M.I.T., University of Chicago, N. Y. City College, and many others. "Excellent introduction . . . remarkably readable, concise, clear, rigorous," *Journal of the American Statistical Association.*

ELEMENTS OF THE THEORY OF FUNCTIONS,
Konrad Knopp

This book provides the student with background for further volumes in this set, or texts on a similar level. Partial contents: foundations, system of complex numbers and the Gaussian plane of numbers, Riemann sphere of numbers, mapping by linear functions, normal forms, the logarithm, the cyclometric functions and binomial series. "Not only for the young student, but also for the student who knows all about what is in it," *Mathematical Journal.* Bibliography. Index. 140pp. 5⅜ x 8.   60154-4 Paperbound $1.50

THEORY OF FUNCTIONS, PART I,
Konrad Knopp

With volume II, this book provides coverage of basic concepts and theorems. Partial contents: numbers and points, functions of a complex variable, integral of a continuous function, Cauchy's integral theorem, Cauchy's integral formulae, series with variable terms, expansion of analytic functions in power series, analytic continuation and complete definition of analytic functions, entire transcendental functions, Laurent expansion, types of singularities. Bibliography. Index. vii + 146pp. 5⅜ x 8.   60156-0 Paperbound $1.50

THEORY OF FUNCTIONS, PART II,
Konrad Knopp

Application and further development of general theory, special topics. Single valued functions. Entire, Weierstrass, Meromorphic functions. Riemann surfaces. Algebraic functions. Analytical configuration, Riemann surface. Bibliography. Index. x + 150pp. 5⅜ x 8.   60157-9 Paperbound $1.50

PROBLEM BOOK IN THE THEORY OF FUNCTIONS, VOLUME 1.
Konrad Knopp

Problems in elementary theory, for use with Knopp's *Theory of Functions,* or any other text, arranged according to increasing difficulty. Fundamental concepts, sequences of numbers and infinite series, complex variable, integral theorems, development in series, conformal mapping. 182 problems. Answers. viii + 126pp. 5⅜ x 8.   60158-7 Paperbound $1.50

PROBLEM BOOK IN THE THEORY OF FUNCTIONS, VOLUME 2,
Konrad Knopp

Advanced theory of functions, to be used either with Knopp's *Theory of Functions,* or any other comparable text. Singularities, entire & meromorphic functions, periodic, analytic, continuation, multiple-valued functions, Riemann surfaces, conformal mapping. Includes a section of additional elementary problems. "The difficult task of selecting from the immense material of the modern theory of functions the problems just within the reach of the beginner is here masterfully accomplished," *Am. Math. Soc.* Answers. 138pp. 5⅜ x 8.
60159-5 Paperbound $1.50

MATHEMATICAL PHYSICS, *D. H. Menzel*
Thorough one-volume treatment of the mathematical techniques vital for
classical mechanics, electromagnetic theory, quantum theory, and relativity.
Written by the Harvard Professor of Astrophysics for junior, senior, and grad-
uate courses, it gives clear explanations of all those aspects of function theory,
vectors, matrices, dyadics, tensors, partial differential equations, etc., necessary
for the understanding of the various physical theories. Electron theory, relativity, and other topics seldom presented appear here in considerable detail.
Scores of definition, conversion factors, dimensional constants, etc. "More
detailed than normal for an advanced text . . . excellent set of sections on
Dyadics, Matrices, and Tensors," *Journal of the Franklin Institute.* Index. 193
problems, with answers. x + 412pp. 5⅜ x 8.   60056-4 Paperbound $2.50

THE THEORY OF SOUND, *Lord Rayleigh*
Most vibrating systems likely to be encountered in practice can be tackled
successfully by the methods set forth by the great Nobel laureate, Lord
Rayleigh. Complete coverage of experimental, mathematical aspects of sound
theory. Partial contents: Harmonic motions, vibrating systems in general, lateral
vibrations of bars, curved plates or shells, applications of Laplace's functions to
acoustical problems, fluid friction, plane vortex-sheet, vibrations of solid bodies,
etc. This is the first inexpensive edition of this great reference and study work.
Bibliography, Historical introduction by R. B. Lindsay. Total of 1040pp. 97
figures. 5⅜ x 8.   60292-3, 60293-1 Two volume set, paperbound $6.00

HYDRODYNAMICS, *Horace Lamb*
Internationally famous complete coverage of standard reference work on
dynamics of liquids & gases. Fundamental theorems, equations, methods, solutions, background, for classical hydrodynamics. Chapters include Equations of
Motion, Integration of Equations in Special Gases, Irrotational Motion, Motion
of Liquid in 2 Dimensions, Motion of Solids through Liquid-Dynamical Theory,
Vortex Motion, Tidal Waves, Surface Waves, Waves of Expansion, Viscosity,
Rotating Masses of Liquids. Excellently planned, arranged; clear, lucid presentation. 6th enlarged, revised edition. Index. Over 900 footnotes, mostly bibliogra-
phical. 119 figures. xv + 738pp. 6⅛ x 9¼.   60256-7 Paperbound $4.00

DYNAMICAL THEORY OF GASES, *James Jeans*
Divided into mathematical and physical chapters for the convenience of those
not expert in mathematics, this volume discusses the mathematical theory of
gas in a steady state, thermodynamics, Boltzmann and Maxwell, kinetic theory,
quantum theory, exponentials, etc. 4th enlarged edition, with new material on
quantum theory, quantum dynamics, etc. Indexes. 28 figures. 444pp. 6⅛ x 9¼.
          60136-6 Paperbound $2.75

THERMODYNAMICS, *Enrico Fermi*
Unabridged reproduction of 1937 edition. Elementary in treatment; remarkable
for clarity, organization. Requires no knowledge of advanced math beyond
calculus, only familiarity with fundamentals of thermometry, calorimetry.
Partial Contents: Thermodynamic systems; First & Second laws of thermodynamics; Entropy; Thermodynamic potentials: phase rule, reversible electric
cell; Gaseous reactions: van't Hoff reaction box, principle of LeChatelier;
Thermodynamics of dilute solutions: osmotic & vapor pressures, boiling &
freezing points; Entropy constant. Index. 25 problems. 24 illustrations. x +
160pp. 5⅜ x 8.   60361-X Paperbound $2.00

NUMERICAL SOLUTIONS OF DIFFERENTIAL EQUATIONS,
  H. Levy & E. A. Baggott
Comprehensive collection of methods for solving ordinary differential equations
of first and higher order. All must pass 2 requirements: easy to grasp and
practical, more rapid than school methods. Partial contents: graphical integration of differential equations, graphical methods for detailed solution. Numerical solution. Simultaneous equations and equations of 2nd and higher orders.
"Should be in the hands of all in research in applied mathematics, teaching,"
Nature. 21 figures. viii + 238pp. 5⅜ x 8.   60168-4 Paperbound $1.85

ELEMENTARY STATISTICS, WITH APPLICATIONS IN MEDICINE AND THE
  BIOLOGICAL SCIENCES, F. E. Croxton
A sound introduction to statistics for anyone in the physical sciences, assuming no prior acquaintance and requiring only a modest knowledge of math.
All basic formulas carefully explained and illustrated; all necessary reference
tables included. From basic terms and concepts, the study proceeds to frequency
distribution, linear, non-linear, and multiple correlation, skewness, kurtosis,
etc. A large section deals with reliability and significance of statistical methods.
Containing concrete examples from medicine and biology, this book will prove
unusually helpful to workers in those fields who increasingly must evaluate,
check, and interpret statistics. Formerly titled "Elementary Statistics with Applications in Medicine." 101 charts. 57 tables. 14 appendices. Index. vi +
376pp. 5⅜ x 8.   60506-X Paperbound $2.25

INTRODUCTION TO SYMBOLIC LOGIC,
  S. Langer
No special knowledge of math required — probably the clearest book ever
written on symbolic logic, suitable for the layman, general scientist, and philos-
opher. You start with simple symbols and advance to a knowledge of the
Boole-Schroeder and Russell-Whitehead systems. Forms, logical structure, classes,
the calculus of propositions, logic of the syllogism, etc. are all covered. "One
of the clearest and simplest introductions," Mathematics Gazette. Second enlarged, revised edition. 368pp. 5⅜ x 8.   60164-1 Paperbound $2.25

A SHORT ACCOUNT OF THE HISTORY OF MATHEMATICS,
  W. W. R. Ball
Most readable non-technical history of mathematics treats lives, discoveries of
every important figure from Egyptian, Phoenician, mathematicians to late 19th
century. Discusses schools of Ionia, Pythagoras, Athens, Cyzicus, Alexandria,
Byzantium, systems of numeration; primitive arithmetic; Middle Ages, Renais-
sance, including Arabs, Bacon, Regiomontanus, Tartaglia, Cardan, Stevinus,
Galileo, Kepler; modern mathematics of Descartes, Pascal, Wallis, Huygens,
Newton, Leibnitz, d'Alembert, Euler, Lambert, Laplace, Legendre, Gauss,
Hermite, Weierstrass, scores more. Index. 25 figures. 546pp. 5⅜ x 8.
          20630-0 Paperbound $2.75

INTRODUCTION TO NONLINEAR DIFFERENTIAL AND INTEGRAL EQUATIONS,
  Harold T. Davis
Aspects of the problem of nonlinear equations, transformations that lead to
equations solvable by classical means, results in special cases, and useful
generalizations. Thorough, but easily followed by mathematically sophisticated
reader who knows little about non-linear equations. 137 problems for student
to solve. xv + 566pp. 5⅜ x 8½.   60971-5 Paperbound $2.75

APPLIED OPTICS AND OPTICAL DESIGN,
A. E. Conrady
With publication of vol. 2, standard work for designers in optics is now complete for first time. Only work of its kind in English; only detailed work for practical designer and self-taught. Requires, for bulk of work, no math above trig. Step-by-step exposition, from fundamental concepts of geometrical, physical optics, to systematic study, design, of almost all types of optical systems. Vol. 1: all ordinary ray-tracing methods; primary aberrations; necessary higher aberration for design of telescopes, low-power microscopes, photographic equipment. Vol. 2: (Completed from author's notes by R. Kingslake, Dir. Optical Design, Eastman Kodak.) Special attention to high-power microscope, anastigmatic photographic objectives. "An indispensable work," *J., Optical Soc. of Amer.* Index. Bibliography. 193 diagrams. 852pp. 6⅛ x 9¼.
60611-2, 60612-0 Two volume set, paperbound $8.00

MECHANICS OF THE GYROSCOPE, THE DYNAMICS OF ROTATION,
R. F. Deimel, Professor of Mechanical Engineering at Stevens Institute of Technology
Elementary general treatment of dynamics of rotation, with special application of gyroscopic phenomena. No knowledge of vectors needed. Velocity of a moving curve, acceleration to a point, general equations of motion, gyroscopic horizon, free gyro, motion of discs, the damped gyro, 103 similar topics. Exercises. 75 figures. 208pp. 5⅜ x 8.
60066-1 Paperbound $1.75

STRENGTH OF MATERIALS,
J. P. Den Hartog
Full, clear treatment of elementary material (tension, torsion, bending, compound stresses, deflection of beams, etc.), plus much advanced material on engineering methods of great practical value: full treatment of the Mohr circle, lucid elementary discussions of the theory of the center of shear and the "Myosotis" method of calculating beam deflections, reinforced concrete, plastic deformations, photoelasticity, etc. In all sections, both general principles and concrete applications are given. Index. 186 figures (160 others in problem section). 350 problems, all with answers. List of formulas. viii + 323pp. 5⅜ x 8.
60755-0 Paperbound $2.50

HYDRAULIC TRANSIENTS,
G. R. Rich
The best text in hydraulics ever printed in English . . . by former Chief Design Engineer for T.V.A. Provides a transition from the basic differential equations of hydraulic transient theory to the arithmetic integration computation required by practicing engineers. Sections cover Water Hammer, Turbine Speed Regulation, Stability of Governing, Water-Hammer Pressures in Pump Discharge Lines, The Differential and Restricted Orifice Surge Tanks, The Normalized Surge Tank Charts of Calame and Gaden, Navigation Locks, Surges in Power Canals—Tidal Harmonics, etc. Revised and enlarged. Author's prefaces. Index. xiv + 409pp. 5⅜ x 8½.
60116-1 Paperbound $2.50

Prices subject to change without notice.

Available at your book dealer or write for free catalogue to Dept. Adsci, Dover Publications, Inc., 180 Varick St., N.Y., N.Y. 10014. Dover publishes more than 150 books each year on science, elementary and advanced mathematics, biology, music, art, literary history, social sciences and other areas.